# The Facility Manager's Handbook

# The Facility Manager's Handbook

By

Joseph F. Gustin

**THE FAIRMONT PRESS, INC.**
Lilburn, Georgia

**MARCEL DEKKER, INC.**
New York and Basel

**Library of Congress Cataloging-in-Publication Data**

Gustin, Joseph F., 1947-
    The facility manager's handbook/by Joseph F. Gustin.
      p. cm.
    I. Includes bibliographical references and index.
    ISBN 0-88173-432-2 (electronic)
      1. Facility management--Handbooks, manuals, etc. 2. Plant
engineering--Handbooks, manuals, etc.  I Title.

    TS184.G87 2002
    658.2--dc21

                                                2002029468




Published by The Fairmont Press, Inc.
700 Indian Trail
Lilburn, GA 30047
tel: 770-925-9388; fax: 770-381-9865
*http://fairmontpress.com*

Printed in the United States of America

10 9 8 7 6 5 4 3 2 1

ISBN 0-88173-432-2

While every effort is made to provide dependable information, the publisher, authors, and editors cannot be held responsible for any errors or omissions.

# Dedication

This book is dedicated to DG.

# Contents

# Preface

Facility management is an art as much as it is a science. If it is the art of integrating people with their physical environment—and it is—then the science of facility management lies in executing its complexities and challenges.

Traditionally, facility management has always dealt with the physical aspects of a building. These "bricks and mortar" issues—HVAC, lighting, electric, plumbing, space allocation, and grounds maintenance continue to be the "nuts and bolts" of facility management.

However, the processes involved in addressing these "mainstays" of facility management have become more complex. The proliferation of regulatory mandates—at the federal, state and local levels, worker compensation issues, increased employee litigation, and violence in the workplace have redefined the role of the facility manager. In its most generic form, the role of the facility manager is often seen as "gatekeeper." In its most specific form, the role of the facility manager must be viewed in its hybrid form—operations manager and compliance officer.

*The Facility Manager's Handbook* addresses these issues. It provides a panoramic view of the process by isolating the key considerations of facility management. These considerations include real estate, space and change management, indoor air quality, emergency preparedness and response planning, communications systems and regulatory mandates.

# *Introduction*

Facility managers are the professionals most responsible for integrating people with their physical environment. As such, facility management is both a people and an environmental issue. The hybrid role of the facility manager as operations manager and compliance officer involves people and productivity, and the costs of managing each. The facility manager must coordinate policies and operations with industry standards and practices, as well as with regulatory mandates.

*The Facility Manager's Handbook* provides a rationale for systematically identifying and evaluating the key areas of practice management. For example, Chapter 1 lays the groundwork for understanding what companies and building owners require in terms of their real estate/property investment. Knowing these requirements, facility managers as the agents of building owners and companies, can work to ensure the continuity and/or expansion of business operations.

Chapter 2 reviews the key elements of space management as those elements are needed to create well-designed, comfortable and compliant work environments.

Chapter 3 addresses change management and the need for communicating change initiatives to employees, occupants and tenants.

Chapter 4 presents an in-depth discussion of indoor air quality. For example, indoor pollutants are not only identified, but their sources are identified as well. Additionally, regulatory standards and guidelines that address indoor air pollutants, with their concomitant health effects and control measures are discussed.

Also included in this chapter is a series of sample indoor air quality concerns and solutions.

*The Facility Manager's Handbook* is unique in its scope. In addition to the core elements of facility management, this book also presents an emergency response model plan outline. This outline, adaptable to a specific site, can be used to satisfy the basic elements of emergency planning as determined by the Occupational Safety and Health Administration.

Also included in this book is the Safety, Emergency Response and Hazard Communication Planning Program. Designed to assist facility managers in developing programming that is compliant with regulatory mandates, this model allows for equity and consistency in implementing policy.

*The Facility Manager's Handbook* focuses upon awareness and practice management. By doing so, it turns the challenges of facility management into opportunities for the facility manager. These opportunities are manifested in an enhanced productivity that aligns itself with business operations.

# Chapter 1
# *Real Estate*

While many facility managers may not be directly involved in real estate transactions, as agents of building/property owners, it is important for facility managers to have a basic understanding of what companies and building owners look for in terms of their investment. This basic understanding helps facility managers ensure the continuity and/or expansion of business operations.

Real estate to a facility manager encompasses a multitude of terms—land, buildings, government regulations—local, state, federal; fire and building codes, standards, price per square foot, renovation, new construction, day-to-day operations, virtual offices, environmental laws, insurance claims, outsourcing, HVAC, landscape—and the list goes on. In a business entity where there is no formal real estate department or where a combined real estate/facility management department exists, it may be the role of the facility manager to oversee various real estate responsibilities.

## GOVERNMENTAL FACTORS

Governmental factors have a significant influence on real estate. Governments, local, regional and national impose taxes on properties and businesses entities. Local governments issue land use controls, such as zoning, to regulate real estate activity; and state and federal governments exert land control through environmental laws. State governments exert influence over real estate through police power, eminent domain, taxation and escheat. Finally, the federal government factors that influence real estate are fiscal policy, monetary policy, secondary mortgage markets and government programs.

## REGULATORY AND COMPLIANCE ISSUES

Important issues that impact upon the real estate responsibilities of the facility manager are the various regulatory and compliance issues that are mandated by law. Specific issues include safety in the workplace—is the workplace safe for employees as outlined in the Occupational Safety and Health Act (OSH). Does the facility meet the physical building requirements set forth by the Americans with Disabilities Act (ADA) which is enforced by the Department of Justice. And does the building/facility comply with the environmental issues as outlined by the Environmental Protection Agency (EPA) in terms of waste disposal, clean air, underground storage tanks, environmental site assessments, Superfund cleanup of existing land spaces, etc.

## ENVIRONMENTAL PROTECTION

The federal government, in protecting wildlife, endangered species and wetlands controls the use of land. Governmental agencies may declare a parcel of land a safe haven for an endangered species or for environmental purposes. However, variances may be requested. If the developer or building owner can substantiate that the land use will benefit the community without serious damage to the environment, consideration may be granted by the governing entity.

### Land Use Controls

Both the state and federal governments use land use controls to protect the environment. Regulations involving the restriction of land use are imposed when environmental concerns exist. A landowner's use of his/her land can conflict with governmental regulations. For example, governmental approval is necessary when land use for a hazardous waste site or some industrial use is requested.

### Public Restrictions

Public restrictions include zoning ordinances, building codes, subdivision regulations, and environmental laws. It can also include the appropriation of private land for public use. In cases where private land is appropriated for public use governmental entities will issue just compensation.

## Zoning Ordinances

Zoning ordinances are local laws that allow for specific land uses in specific areas. For example, certain areas of the community are zoned commercial, industrial, light industrial, residential, agricultural, rural, etc. The activities that occur in each of these specific areas are regulated by the governmental entity to protect the public.

Zoning laws usually include minimum land size, building height limits, setback and size of building/facility for each type of land use in the zoned area. For example, lot size for a commercial/industrial zone differs from the lot size in a residential zone. Zoning laws also include limits on off-street parking, landscaping, and outdoor lighting, etc.

## Building Codes

Building codes set construction standards. Depending upon the local regulations builders are required to use particular methods and materials. Local building codes set regulations for fire codes, plumbing codes, electrical codes, etc. In Ohio, for example, state law sets minimum building standards, but local governments can mandate additional requirements.

Permits are issued to enforce building codes. Plans for building or remodeling a facility must be submitted to the local building department for approval. A permit is then issued if the plans comply with the codes. Inspections by local governmental agencies may be periodically conducted throughout the construction, renovation and completion phases. Stop work orders can be issued if there is non-compliance with applicable building codes. Fines and injunctions are used to enforce building codes.

In addition to building code inspections many governmental entities conduct routine fire and health inspections for most commercial, non-residential buildings. Additionally, when more stringent standards are imposed, property owners may be required to fit their building to the new code standards. Again, fines and injunctions are used to enforce building codes.


## ENVIRONMENTAL PROTECTION— FEDERAL ENVIRONMENTAL LAWS

Environmental laws are laws enacted by federal and state governments to protect the country's land, air and water. These laws are designed to keep the land, air and water clean, and to promote conservation of land, air, water and natural resources. Here's a brief look at some of the most important federal and state environmental laws.

### The Clean Air Act

The Clean Air Act regulates air emissions from area, stationary, and mobile sources. This comprehensive federal law authorizes the U.S. Environmental Protection Agency (EPA) to establish National Ambient Air Quality Standards (NAAQS) to protect public health and the environment. The goal of the Act was to establish and achieve National Ambient Air Quality Standards (NAAQS) in every state by 1975. The setting of maximum pollutant standards was coupled with directing the states to develop State Implementation Plans (SIP's). States are required to prepare a state implementation plan (SIP) for meeting national air quality standards of the EPA. States are authorized to stop projects that interfere with clean air objectives that are applicable to appropriate industrial sources in each state.

The Act was amended in 1977 primarily to set new goals (dates) for achieving attainment of NAAQS since many areas of the country had failed to meet the deadlines. The 1990 amendments to the Clean Air Act in large part were intended to meet non-addressed or insufficiently addressed problems such as acid rain, ground-level ozone, stratospheric ozone depletion, and air toxins.

**The Clean Water Act**

The Clean Water Act is a 1977 amendment to the Federal Water Pollution Control Act of 1972. The Federal Water Pollution Act regulates the discharge of pollutants into the waters of the United States. This law gives the Environmental Protection Agency (EPA) the authority to set effluent standards on an industry basis (technology-based) and set water quality standards for all contaminants in surface waters. The Clean Water Act (CWA), an amendment to the Federal Water Pollution Control Act, makes it unlawful for any person to discharge any pollutant from a point source into navigable waters unless a permit is obtained. The CWA was reauthorized in 1987 focusing on toxic pollutants and toxic substances, authorizing citizen suit provisions, and funding sewage treatment plants.

The Clean Water Act also dictates that the adequacy of treatment facilities be considered before allowing new construction, and encourages localities to look at new technology and alternatives to wastewater plants.

The National Environmental Policy Act (NEPA) requires federal agencies to prepare an environmental impact statement (EIS) for any project that would have a significant effect on the environment. An EIS details a project's impact on energy use, sewage systems, drainage, factories, schools and other environmental, economic, and social areas. NEPA applies to all federal development projects, such as dams or highways. It also applies to private projects that require a license/permit from a federal agency, or the issuance of a federal loan.

**The National Environmental Policy Act (NEPA)** established a broad national framework for protecting the environment. When governmental facilities such as airports, buildings, military complexes, highways, parkland purchases, etc. are proposed, the tenets of NEPA are then referenced. The basic policy of NEPA is to assure that proper considerations will be given to the environment by all branches of the government before any major federal action that significantly affects the environment is undertaken.

**Occupational Safety and Health Act**

The Occupational and Safety Health Act was passed to ensure worker and workplace safety. This act provided that employers provide their workers a place of employment free from recognized hazards to safety and health, such as exposure to toxic chemicals, excessive noise levels, mechanical dangers, heat or cold stress, or unsanitary conditions.

In order to establish standards for workplace health and safety, the Act also created the National Institute for Occupational Safety and Health (NIOSH) as the research institution for the Occupational Safety and Health Administration (OSHA). OSHA is a division of the U.S. Department of Labor that oversees the administration of the Act and enforces standards in all 50 states.

**Comprehensive Environmental Response, Compensation, and Liability Act**

The Comprehensive Environmental Response, Compensation, and Liability Act (CERCLA) is better known as the Superfund. This act mandates a tax on the chemical and petroleum industries to provide a trust fund for cleaning up abandoned or uncontrolled waste sites. It also mandates broad Federal authority to respond to hazardous substance releases or threatened hazardous substance releases that may endanger public health or the environment. CERCLA:

- Established prohibitions and requirements concerning closed and abandoned hazardous waste sites;

- Provided for liability of persons responsible for releases of hazardous waste at these sites; and

- Established a trust fund to provide for cleanup when no responsible party could be identified.

The law also authorizes two kinds of response actions:

- Short-term removals, where actions may be taken to address releases or threatened releases requiring prompt response.

- Long-term remedial response actions, that permanently and significantly reduce the dangers associated with the releases or threats of releases of hazardous substances that are serious, but not immediately life threatening. These actions can be conducted only at sites listed on EPA's National Priorities List (NPL).

**Superfund Amendments and Reauthorization Act**

The Superfund Amendments and The Authorization Act of 1986 reauthorized CERCLA to continue cleanup activities around the country. Several site-specific amendments, definitions clarifications, and technical requirements were added to the legislation, including additional enforcement authorities. SARA:

- Stressed the importance of permanent remedies and innovative treatment technologies in cleaning up hazardous waste sites;

- Required Superfund actions to consider the standards and requirements found in other State and Federal environmental laws and regulations;

- Provided new enforcement authorities and settlement tools;

- Increased State involvement in every phase of the Superfund program;

- Increased the focus on human health problems posed by hazardous waste sites;

- Encouraged greater citizen participation in making decisions on how sites should be cleaned up; and

- Increased the size of the trust fund to $8.5 billion.

SARA also required the Environmental Protection Agency to revise the Hazard Ranking System (HRS) to ensure that it accurately assessed the relative degree of risk to human health and the environment posed by uncontrolled hazardous waste sites that may be place on the National Priorities List (NPL).

Title III of SARA also authorized the Emergency Planning and Community Right-to-Know Act (EPCRA).

In Region 5, for example, SARA is administered by the Superfund Division.

## Freedom of Information Act

The Freedom of Information Act (FOIA) provides specifically that "any person" can make requests for government information. Citizens who make requests are not required to identify themselves or explain why they want the information they have requested. The position of Congress in passing FOIA was that the workings of government are "for and by the people" and that the benefits of government information should be made available to everyone.

All branches of the Federal government must adhere to the provisions of FOIA with certain restrictions for work in progress (early drafts), enforcement of confidential information, classified documents, and national security information.

## Endangered Species Act

This Act provides a program for the conservation of threatened and endangered plants and animals and the habitats in which they are found. The U.S. Fish and Wildlife service (FWS) of the Department of the Interior maintains the list of 632 endangered species (326 are plants) and 190 threatened species (78 are plants).

Species include birds, insects, fish, reptiles, mammals, crustaceans, flowers, grasses, and trees. Anyone can petition FWS to include a species on this list. The law prohibits any action, administrative or real, that results in a "Taking" of a listed species, or adversely affects the habitat. Likewise, import, export, interstate, and foreign commerce of listed species are prohibited.

EPA also registers pesticides. Pesticides can pose the risk of adverse effects on endangered species as well as on the environment. EPA can issue emergency suspensions of certain pesticides to cancel or restrict their use if an endangered species will be adversely affected.

## Oil Pollution Act of 1990

The Oil Pollution Act (OPA) of 1990 strengthened EPA's ability to prevent and respond to catastrophic oil spills. EPA publishes regulations for above-ground storage facilities; the Coast Guard publishes one for oil tankers The OPA requires oil stage facilities and vessels to submit to the Federal government plans detailing how they will respond to large discharges. A tax on oil financed a "trust fund" to clean up spills when the responsible party is incapable or unwilling to do so. The OPA also requires the development of Area Contingency Plans to prepare and plan for oil spill response on a regional scale.

## Pollution Prevention Act

The Pollution Prevention Act focused on source reduction of pollution rather than waste management or pollution control.

Pollution prevention also includes practices that increase efficiency in energy use, water use, etc. Conservation of resources is the basis of this act and focuses upon recycling, source reduction, and sustainable agriculture.

## Resource Conservation and Recovery Act (RCRA)

RCRA is the act the enables the EPA to control hazardous waste in every cycle or stage from the generation, transportation, treatment, storage, and disposal of hazardous waste. RCRA also mandates a framework for the management of non-hazardous wastes.

The 1986 amendments to RCRA enabled EPA to address environmental problems that could result from underground tanks storing petroleum and other hazardous substances. RCRA focuses only on active and future facilities and does not address abandoned or historical sites.

**The Federal Hazardous and Solid Waste Amendments (HSWA)**, an amendment to the RCRA, requires the "phasing out" of the practice of land disposal of hazardous waste. Increased enforcement authority for EPA, stringent hazardous waste management standards and a comprehensive underground storage tank program are some of the major mandates of this amendment.

**Toxic Substances Control Act**

The Toxic Substances Control Act (TSCA) of 1976 gives the EPA the ability to track the 75,000 industrial chemicals currently produced or imported into the United States. EPA repeatedly screens these chemicals and can require reporting or testing of those that may pose an environmental or human-health hazard. EPA can ban the manufacture and import of those chemicals that pose an unreasonable risk.

Also, EPA tracks thousands of new chemicals that industry develops each year with either unknown or dangerous characteristics. EPA can control these chemicals to protect human health and the environment. TSCA supplements other Federal statutes, including The Clean Air Act and The Toxic Release Inventory under EPCRA.

## STATES' ENVIRONMENTAL LAWS

To comply with federal requirements, many states, including Ohio, passed air and air pollution control laws, and set up their own versions of the EPA. The Ohio Environmental Protection Agency (OEPA), for example, passed its own version of air and water pollution control laws and administers those laws. An OEPA permit is required for any large discharge of pollutants into the water or air. OEPA's grant or denial of a permit can be appealed to the state Environmental Board of Review. OEPA does not require an EIS, but does review any federal EIS prepared for an Ohio project. Check with your own state for their EPA rules.

## LEASES

A common real estate transaction is leasing a place to live or work. Facility managers, while not necessarily involved in the direct purchase of property or leasing, do need a basic understanding of the real estate function as it relates to their facility and landlord/tenant relationships, since they may be asked to provide input or to provide consultative input to building owners into the rental and/or sale of their property.

Leases are contracts in which one party (landlord or lessee) gives to another party (tenant, or lessee) the use and possession of lands, buildings, etc. for a specified time and for specific payments. The lease, then, is the conveyance of a leasehold estate from the fee owner to a tenant. Commercial leases consist of a gross lease, net lease, percentage lease and land lease (or ground lease).

## Gross Lease

In a gross lease the building/property owner or landlord pays the property taxes, mortgage payments, insurance, etc., while the tenant pays the utilities. A gross lease can provide for future rent increases tied to inflation or taxes. A gross lease is usually used for tenants such as lawyers, doctors and executives.

## Net Lease

In a net lease the tenant pays the property taxes, mortgage payments, insurance, etc., as well as all utilities, in addition to a monthly rent payment. A net lease is a variation of the gross lease. However, increased costs for rent are shifted from the landlord to the tenants who pay a proportional amount of the property taxes, insurance and sometimes maintenance costs.

## Percentage Lease

In a percentage lease the tenant pays a percentage of their gross sales to the landlord, often in addition to a fixed monthly rental payment. Utilities may or may not be included. A percentage lease usually is used by retail tenants.

## Land Lease

In a land lease (or ground lease) a tenant leases only the land from the landlord, but the tenant actually owns the building. A land lease is usually executed when the land is vacant. The owner of the building can sell the building at any time and have the buyer assume the land lease. The landlord also can sell the land subject to the lease of the building owner. A land lease usually lasts the life expectancy of the building, but can last as long as 99 years.

## OWNERSHIPS

Ownership can also be further categorized into ownership by associations. Associations may be not-for-profit, a business, a corporation or a partnership.

Corporations

A corporation is a group of people who get a charter granting them, as a body, certain legal powers, rights, privileges and liabilities of an individual. These legal powers, however, are distinct from those individuals that make up the group. A corporation can buy, sell, and inherit property. It is also a legal entity in which individuals hold shares of stock. These shareholders, however, do not own the property of the corporation. The shareholders are also not liable personally for its debts. To sell the property, the corporation must authorize the purchase or sale of the real estate.

General Partnerships

A partnership is an association of two or more people who contribute money or property to carry on a joint business; they are co-owners of the business, and, as such, share in its profits and losses. The partners own the property of the partnership as tenants in partnership; it is a co-ownership in which each partner has an equal and undivided interest in the property. The title of the property can be held in the partnerships' name. In Ohio, for example, the county in which the business is located issues a business certificate. This certificate lists the name of the partnership, the location of the office and the location of the real property along with the names and addresses of all the partners. In a general partnership, the deed to any land held in the name of the partnership needs only the name of one partner as the grantee and the property itself can be sold or mortgaged by any of the partners.

## Limited Partnerships

A limited partnership is also an association of two or more persons who contribute money or property to carry on a joint business; they are also co-owners of the business, and, as such, share in its profits and losses. A limited partnership has one or more general partners and one or more limited partners. Limited partners, however, do not have any input into partnership matters.

## REITs

A Real Estate Investment Trust (REIT) is a real estate investment business organized as a trust. It has at least 100 investors. The trustees, one or more persons, manage the property for the investors. The title of the property is vested in the name of the trustees. These trustees are bound to the requirements of the trust document. The investors enforce the performance of the trust, but do not have a legal interest in the property.

## PROPERTY APPRAISALS

Appraisals are performed to determine loans, insurance and taxes. A property appraisal is an estimate or opinion of value of a specific piece of property at a specific place in time. The appraisal, then, is an estimate or opinion of value supported by objective evidence and data. To determine the value of a property, the appraiser

(1)  calculates the cost to reproduce or replace the existing structure;

(2)  subtracts from the cost estimate any loss in value because of depreciation; and

(3)  adds that value of the site alone to the depreciated cost figure.

Land value, however, is appraised separately and then added to the depreciated construction cost of the structure. Site value is determined by the location and improvements (exclusive of buildings) of the specific property, or site. The appraiser then finds recently sold comparable properties. Sale price adjustments are made to account for any significant differences between the comparable and the specific property. The value of the property is then based upon an estimate of the site values based on the adjusted values of the comparables.

**Sources**
Hondros College, *Ohio Real Estate Principles and Practices*, Rev. ed., Columbus: Lash Publications, 1996, 1997

# Chapter 2
# *Space Management*

**W**hen considering the space management and workplace planning needs of the modern office, space management specialists, designers and facility managers must take into account several key factors. First, attention must be paid to creating well-designed, comfortable and compliant working environments.

To achieve the goal of a well-designed, comfortable, compliant and efficient working environment, facility managers should, along with their planning teams, identify the needs of the organization in terms of the management of space and incorporate these needs into the plan. One of the key considerations in the management of space is Problem Identification.


## PROBLEM IDENTIFICATION

When identifying the key issues of space management in the facility/organization, facility managers, along with their planning team should rely on the elements of a good plan—i.e., a defined purpose/objective, the scope and, finally, the implementation of the plan. The key components of the plan are the tenant/occupants, space and technology. A consistent approach, a functional/flexible space environment and long-term savings are the benefits of good planning. Organizational requirements, technological requirements, HVAC requirements, lighting, security, ADA signage and accommodations, and intended benefits are among many of the factors that should be incorporated into the plan. Also needed in the plan is a long-range view of the future uses, organizational and technological requirements along with the present and future management of the plan.

## The Planning Phase

The planning phase is the heart of the project. In this phase, the organization needs to examine its mission, the reasons for the project and the expected outcomes that will later measure the success of the plan. Thought should be given to the present ways of doing business and the real and probable future changes in business practices. The organizational requirements for the successful implementation of the plan include flexibility, minimization of barriers and encouragement of teamwork and interaction.

During the planning phase, strategic and master plans for the future will be developed. Core business issues, customer relations, finance, administration and information technology are among the basis for forming the plan. Assembling a team to develop plans is a first step in the planning phase. The planning committee should consist of in-house personnel and the appropriate professional consultants.

## The Planning Team

The planning team should include selected representatives from in-house personnel that represent the makeup of the organization. These personnel should be able to communicate the organization's goals and ideals to both those planning and those implementing the project. Management, financial officials and the occupants of the building—those that have a vested interest in the outcome should also comprise the planning team. Worker interaction should be part of the goals along with ideas about the types of technology. Representatives from the departments of design, facility management, human resources, finance, marketing, information technology and communication, and most importantly, space users/tenants/occupants themselves, and, if applicable, union representatives will make up the planning committee.

## SPACE PLANNING CONCEPTS/CONSIDERATIONS

A second key element of the planning phase addresses the management of space and workplace planning, itself. Integral to these space-planning concepts and which should be considered include alternative office spaces. Alternative office spaces determine how, when and where people work; once this is done, a match of those needs to the range of work environment solutions can be made. For example, if a person may not be required to come in to the office location daily, then an off-site location can be used. Alternative office space can also include innovative on-premise arrangements. Some of these office options includes: hoteling, moteling, shared space, caves and commons/privacy space, free address, home base, relief space, team or group address/co-location, and conference/multimedia spaces. Off-premises office options include: teleworking/telecommuting, satellite offices and telecenters and virtual office.

**On-premise office utilization strategies** use existing office space:

- Hoteling—Employees call to reserve workspace in the main office where there are fewer offices than staff.

- Moteling—Employees check in upon arrival and are assigned a workspace with no advance "reservations."

- Shared Space—Two or more employees use a single workspace permanently assigned to them. Employees must work together to determine when they each can use the space.

- Caves and Commons/Privacy Space—Individual workspace is provided for tasks requiring high levels of concentration, and common, community space areas are provided for team activities, like gatherings and conversations.

- Free Address—A mix of unassigned private and open plan offices and team and retreat areas are combined into one large integrated space.

- Home Bases—An arrangement where an employee has a permanent office or workstation, as well as a workspace with another group where they work on a specific project assignment.

- Relief Space—Here, employees can interact with other employees to generate spontaneous or creative problem solving.

- Team or Group Address/Co-Location—This work environment is designated for use by a specific project team for the duration of a project.

- Conference/Multimedia Spaces—Special communication and presentation facilities that are provided in a separate space.

Where an employee is not required to come in to the office everyday, the off-premise strategy is addressed. This particular "officing-strategy" includes:

- Teleworking/Telecommuting—Employees work at home, at an alternate work facility, or in a "virtual" environment at least part-time.

- Satellite Offices and Telecenters—Alternate work facilities are located nearer to employees' homes. Satellite offices are typically operated by and for a single employer, while telecenters operate independently and are used by numerous employers.

- Virtual Office—Employees are equipped with the tools, technology, and skills to perform their jobs from anywhere, including home, office, customer location, or in transit.

In a study commissioned by the General Services Administration (GSA), that agency noted the most common space planning concepts in use today. These concepts include assigned, or territorial, workspaces and the unassigned (non-territorial, just-in-time) workspaces.

Examples of assigned or territorial workspaces are the
  Vast Open Plan
  Cluster Open Plan
  Closed Offices, Shared Closed Offices, and Combi-Offices
  Combination Open and Closed Offices
  Caves and Commons
  Mobile Workstation
  Portable Office

Examples of unassigned (non-territorial, just-in-time) workspaces:
  Free Address
  Hoteling
  Red Carpet Club Offices

The GSA Public Buildings Service has identified major office space concepts:

**Privacy Spaces**: Where employees can work and concentrate without constant interruptions but still have access to voice and data communications.

**Team Space**: Dedicated spaces, available either on an ad hoc daily basis or for longer periods, which allow group tasks to be done.

**Universal Planning**: A common size or module that fits most uses. This design tool can accommodate fixed and open design concepts.

**Community Space Areas**: Areas set up for informal social gathering and conversation (e.g., eating areas, kitchen and coffee bar).

**Conference Room and Multimedia Spaces**: Enclosed space with separate capability for a television/video monitor, slide projector, etc.

**Hoteling and Moteling**: Non-assigning workstations that can be individually reserved for use.

**Aesthetics**: Use of color, materials and layout to create a pleasing, stimulating, appropriate, and professional environment.

**Views and Vistas**: Providing the maximum views to the outside by leaving the window walls unblocked.

**Ergonomics**: The entire environment, not just furniture, designed and provided to benefit the health and comfort of the employees.

**Signage/Wayfinding**: Aids provided to help locate and identify offices and create a sense of place.

**System Modularity**: A regular, interchangeable system that allows for maximum flexibility and reconfiguration of the space to meet changing functional needs and number of staff.

**System Redundancy, Durability, and Maintainability**: HVAC, lighting, power, security, and telecommunications systems with back-up capabilities to ensure minimal loss of service.

**Artificial Lighting**: Appropriate levels and types (direct, indirect, and task) of lighting allowing employees to control levels of light as necessary.

**Daylighting**: Natural daylight supplied to the greatest number of people by leaving the window walls unblocked or using translucent partitions or windows where closed spaces are necessary.

**Air Quality**: Allowing workers to adjust the space for thermal control to meet personal and team comfort levels for fresh air, thermal control, humidity, and odor control.

**Interior Landscaping**: Natural plants and vegetation, helping to produce healthy air and create an attractive working environment.

**Acoustics/Noise Reduction**: Providing a suitably quiet work environment or quiet "spaces" within the general office space— sound masking (white or pink noise) can also be utilized.

**Communications**: Conditions permitting easy communications among workers, while allowing simultaneous access to data.

**Safety/Security**: Adherence to health and fire safety codes and physical and technical security precautions so employees feel safe and comfortable.

For each of these space planning concepts, the planning committee should consider:

- How business goals are supported.
- Impact upon business operations.
- How daily work routines will be affected.

## FURNITURE

Furniture selection is key to the tenant/occupant issue and the organizations/facility's image or corporate style. The furniture selected should meet the needs of the tenant/occupant/employee and also project and enhance the company's/organization's corporate image. The major factors to consider when selecting furniture are:

- **Task support**: Is the furniture appropriate for the type of work tasks?

- **Flexibility**: Does it accommodate the physical and personal needs of the users? Does it also accommodate the present needs as well as anticipated needs of the workforce?

- **Ergonomics**: Are healthful working conditions supported?

- **Space requirements**: Does it fit the purposes for which it is intended?

- **Value**: Does the cost justify the utility?

- **Technological accommodation**: Will it support the present workstations and peripherals as well as the future uses?

- **Image/Aesthetics**: Is the corporate image enhanced? Is the space utilitarian and functional as well as comfortable?

- **Accessibility**: Is the workstation/furniture adaptable and ADA compliant?

## TENANT/OCCUPANT ISSUES

Another key element in space planning is the tenant/occupant issue—are the needs of the individual tenant/occupant met in terms of that occupant's work processes. Workplace alternative, i.e., alternative officing, either on-premise or off-premise, should be explored and a plan for organizational change should be developed and managed. The individual's needs for personal comfort in terms of indoor air quality, ergonomically designed furniture and lighting and space requirements should be satisfied. For example, the space needs of a secretary or file clerk are sometimes greater than their supervisor. Spatial characteristics should be understood by the tenant/occupant and these characteristics should be addressed. Finally, suitable space planning concepts should be used.

## TECHNOLOGY

Another factor to consider in space planning is technology. The tools that are needed by the tenant/occupant should be in place—telephone/modem outlets, power outlets, computer outlets, defined workspaces, etc. The technology that is/will be used should support the organization and its work practices and, at the same time, be able to accommodate future change. The space plan should be cost effective now and in the future. And, finally, this technology should use suitable procurement and maintenance methods.

**Sources:**

U.S. General Services Administration, Office of Government wide Policy, Office of Real Property, *The Integrated Workplace, A Comprehensive Approach to Developing Workspace*, May, 1991. Second Printing, April, 2000.

# Chapter 3
# Change Management

Change management is a loosely defined term that refers to a broad array of activities and initiatives that occur in the workplace. As such, in order to be effective, a change management program must integrate those program elements that address any of the variety of elements: communication, training and testing, program planning, market analysis and implementation of new policies and procedures.

## CHANGE MANAGEMENT STRATEGY

Adjusting to the changes in the workplace is the focus of a change management program. A good change management strategy includes:

- Communication
- Discussion
- Involvement

### Communication

In any business setting, a good change management program must include communication. Communication is at the core of the change management program. The change itself should be communicated to the tenant/occupant/employee along with the hows, whys and wherefores. Prerequisites to an effective change management approach are:

- Definition
- Explanation
- Progression

**Definition**—Defining change is an essential aspect of a change management strategy. For example, a definition of the major goal of the project must be determined. Is the goal to support current business needs, or is the goal intended to improve customer service, or attract and retain employees? The employee/tenant/occupant should know the reason for the change.

**Explanation**—explaining how it will occur and why it needs to occur, as well as describing when it well occur is another important part of the communication process.

**Progression**—providing the timely information on the various stages of implementation are prerequisites to an effective change management approach.

## Discussion

Employees/occupants/tenants may be anxious, fearful, etc. about the change that is about to occur. Discussing what their role and integration into the change should alleviate the concerns of the employee/occupant/tenant. With employees, for example, discussing such issues as career development, performance assessment, technical skills and work and family issues should alleviate some of their concerns. For tenant/occupants, issues such as how the change will impact upon their operations are an issue that must be discussed. In fact, bringing the tenants/occupants into the change process is one of the most effective ways to secure their commitment and support. A planning committee comprised of the building owners/executives, company manager and "rank-and-file" serves an additional purpose. New suggestions and new ideas for implementing change can be generated from such a group.

## Involvement

As discussed previously, involving employees/tenants/occupants in the planning and design; communicating the change, information about the change including updating the employee/ tenant/occupant about the change; and, finally, soliciting their input, should provide greater support and commitment on their part to the change.

## Developing a Communications Strategy

A second aspect to the change management strategy includes developing a communications strategy.

Facility managers can use a number of methods to plan and implement a communications strategy for all aspects of their change management initiative. Rooted in the principles of marketing, these methods involve:

- Market analysis.
- Program planning.
- Promotion.

Each of these factors, in turn, involves:

- Accurate awareness of a company's change management needs.
- Adapting to an organizational mandate.
- Organized communications systems.
- Cost and quality control.

## Market Analysis

Effective "market analysis" begins by focusing on the company's current operating environment to identify safety and health hazards and potential hazards that are present in the workplace. Market analysis provides the necessary information for addressing hazards, vis-à-vis program planning. Good market research allows companies to make the most efficient use of their resources and measures success in safety and health communications

Market analysis identifies the communication need.

## Program Planning

Program planning is product development—the process of bringing staff, resources (internal and external) and capacity together to meet the company's safety and health needs, as well as ensure compliance with regulatory mandates. Like facility planning, effective program planning (or product development) expedites bringing new programs, products and services on board, while assuring the cost-efficiency of the result.

Program planning develops the structure and systems to deliver communications and satisfy the facility's change management requirements.

### Promotion

The promotional aspects of marketing serve as a logical follow-up to program planning. These promotional aspects facilitate a practical understanding of the company's change management program.

Promotion is the active process of communicating safety policies, training and other important safety and health information to the building occupants.


## EFFECTIVENESS OF THE COMMUNICATIONS STRATEGY

The effectiveness of the communications strategy is determined by how well it facilitates a practical understanding of the change management initiative, as well as the responsibilities that each employee assumes.

The specific criteria used to evaluate effectiveness include the right combination of structure, systems and staff skills.

### Structure

By nature, market and program planning cross all organizational lines and require the input and commitment of all employee levels. Analysis, then, becomes the function of all employees, with each person assuming responsibilities for change management and acting upon those responsibilities.

Effective program planning, like facilities planning, requires a skilled integrator—the facility manager, who is in charge and responsible for the results of the company's change management initiative. Facility managers must work with the Change Planning Committee, the organizationally sanctioned group that has the authority to make and enforce decisions.

## Systems

All of the company's operating systems must be utilized to implement marketing and program planning strategies, including market analysis, budgeting, financial planning, legal resources and human resources.

Managers and supervisors are responsible for ensuring the successful change in their respective divisions and departments. Facility managers have the prime responsibility for the company's overall change management initiative. All managers and supervisors should incorporate practical considerations in their specific communications strategies.

## Staff Skills

Staff skills, including analysis, consensus-building and process management must be utilized to fully implement the company's change management communication strategy. The practical considerations for doing so are described below.

## MAKING THE STRATEGY WORK

Making any change management program work effectively involves a number of communications and management techniques. They include:

- Developing "open" lines of communication.
- Communication incentive.
- Management techniques.

### Developing Open Lines of Communication

Open communications lines provide for a "free" flow of information, ideas and problem-solving techniques. Developing these open lines is essential to the change management strategy.

Examples include meetings, change management committees, company newsletters and "change" hotlines.

*Meetings*

Multilevel meetings provide the vehicles for informing employees/tenants/occupants at all levels of the change issues and the current developments. Regularly scheduled peer-level meetings, as well as manager-employee meetings, give everyone in the company an opportunity to exchange ideas, concerns, and the opportunity to provide input into company direction and policy.

*"Change" Committees*

A formal change management committee, comprised of management and "rank and file" representatives should be instituted, and meet on a regularly scheduled basis to discuss problems and concerns.

Committee actions, recommendations, etc., should be "publicized" so that all employees are updated on a company's safety and health efforts.

*Company Newsletters*

Every newsletter should contain a "change section" that, again, keeps employees apprised of the organization's change management efforts.

Recognizing divisions, departments and individual employees for their own particular "change initiatives/implementations" keeps the change issue in the forefront, while giving everyone an opportunity to "share the spotlight."

*"Change" Hotlines*

To ensure effective communication bases with corporate offices, multi-site companies can consider establishing a hotline, so that any serious problems are addressed immediately.

**Communication Incentives**

Communication incentives are non-financial opportunities that can be used to let employees know their worth to the organization. Some of these incentives include "change" recognition awards programs and/or luncheons where employees are publicly acknowledged for their safety efforts. Employee profiles in annual reports, shareholder notices, or client mailings are other techniques that employers may want to consider in recognizing the change-committed employee.

## Sources

Gustin, Joseph F., *Safety Management, A Guide for Facility Managers*, UpWord Publishing, New York, 1996.

U.S. General Services Administration, Office of Governmentwide Policy, Office of Real Property, *The Integrated Workplace, A Comprehensive Approach to Developing Workspace*, May, 1991. Second Printing, April, 2000.

# Chapter 4
# *Indoor Air Quality*

Good indoor air is essential for the efficient operation of a facility. It promotes workplace productivity. With good indoor air quality, odors, dust and contaminants are kept to a minimum. The circulating air prevents drafts and stuffiness. The temperature is comfortable during all the seasons. Thus employee/occupants feel more comfortable, healthy and productive.

Good Indoor Air Quality also promotes building marketability. Potential buyers or lessees see advantages to owning or leasing a well-maintained and a well-managed building.

## BASICS OF INDOOR AIR QUALITY

According to the Environmental Protection Agency, good air quality is part of a healthy indoor environment. The definition of good indoor air quality, according to that agency, includes:

- control of airborne contaminants
- maintenance of acceptable temperature and relative humidity
- introduction and distribution of adequately ventilated air

On the other hand, poor Indoor Air Quality leads to health problems such as cough, eye irritation, headache, and allergic reactions (in extreme cases, the EPA has noted, life-threatening conditions such as Legionnaire's disease and carbon monoxide poisoning could occur).

In these cases, productivity is reduced and a marked increase in absenteeism can occur.

From a marketability standpoint, poor indoor air quality can lead to a number of less than positive consequences, including:

- accelerating depreciation of furnishings and equipment;
- creating publicity that could put properties at a competitive disadvantage;
- opening potential liability problems (Insurance policies tend to exclude pollution-related claims).

Good indoor air quality starts with a commitment from the top. Whether it is the building owner, a company's senior management team or the facility manager, the person who develops policy and assigns staff responsibilities must ensure that sound indoor air quality practices are developed and implemented.

Such practices must include developing a communications policy that addresses (1). equipment maintenance, (2). energy costs and (3). complaint response.

### Communications Policy

An effective communications policy enhances the effort to diagnose and correct problems. Indoor air quality problems can be prevented if staff and building occupants are made aware of how their activities affect air quality. Depending upon the size or use of the facility, a health and safety committee or a joint management-tenant IAQ task force can establish a list of indoor air quality guidelines and communicate these guidelines to promote good working conditions. This group can help to:

- disseminate information about indoor air quality,

- bring potential problems to the attention of building staff and management;

- and foster a sense of shared responsibility for maintaining a safe and comfortable indoor environment.

### (1). Equipment Problems
The facility staff should be alert to malfunctioning equipment

or accidents that can produce indoor air quality problems. It is the staff that can play a definitive role in identifying and preventing potential problems as well as averting problems.

*(2).  Minimizing Energy Costs*

Some efforts to minimize energy costs can contribute to poor indoor air quality. For example, reducing temperature to save heating costs can impact ventilation systems and increase moisture problems.

*(3).  Responding to Indoor Air Quality Complaints*

Many indoor air quality problems and/or complaints can be solved by the company's facility staff. Once a problem has been identified, facility managers and their staff can begin to correct the problem. Air quality checklists, like the one described below, are helpful in identifying problem areas, sources of contaminants, as well as potential problem areas. These checklists will help the building's facility staff monitor indoor air quality.


## ELEMENTS OF INDOOR AIR QUALITY

The indoor environment of a building is a combination of the following:

- The site.

- The building's climate system including the original design, and any later modifications in the structure and mechanical systems.

- Construction techniques.

- Contaminant sources including building materials and furnishings, moisture, processes and activities within the building.

- Outdoor sources and

- The building occupants.

| Item | Date begun or or completed (as applicable) | Responsible person (name, telephone) | Location ("NA" if the item is not applicable) |
|---|---|---|---|
| **IAQ PROFILE** | | | |
| Collect and Review existing records | | | |
| HVAC design data, operating instructions and manuals | | | |
| HVAC maintenance and calibration records, testing and balancing reports | | | |
| Inventory of locations where occupancy, equipment, or building use has changed | | | |
| Inventory of complaint locations | | | |
| **Conduct a Walkthrough Inspection of the Building** | | | |
| List of responsible staff and/or contractors, evidence of training, and job descriptions | | | |
| Identification of areas where positive or negative pressure should be maintained. | | | |
| Record of locations that need monitoring or correction | | | |
| **Collect Detailed Information** | | | |
| Inventory of HVAC system components needing repair, adjustment, or replacement. | | | |
| Record of control settings and operating schedules | | | |
| Plan showing airflow directions or pressure differentials in significant areas. | | | |
| Inventory of significant pollutant sources and their locations | | | |
| MSDSs for supplies and hazardous substances that are stored or used in the building | | | |
| Zone/Room Record | | | |

Source: U.S. Environmental Protection Agency

## Figure 4-1. IAQ Management Checklist (Sample)

According to the United States Environmental Protection Agency, indoor air quality problems result from four elements:

- **Source**: there is a source of contamination or discomfort indoors, outdoors, or within the mechanical systems of the building.

- **HVAC**: the HVAC system is not able to control existing air contaminants and comfortable temperature and humidity conditions.

- **Pathways**: one or more pollutant pathway connects the pollutant source to the occupants and a driving force exists to move pollutants along the pathway.

- **Occupants**: building occupants are present.

The EPA lists the basic approaches to mitigating indoor air quality problems and concludes that the successful mitigations include a combination of the following techniques:

- Control of pollutant sources;
- Modifications to the ventilation system;
- Air cleaning; and
- Control of exposure to occupants.

Contributing to indoor air quality are such diverse elements as new office equipment, new and existing furnishings, housekeeping activities, room dividers, and the personal activities of the building occupants. Large and multi-use buildings such as apartment buildings, hospitals, schools, shopping malls and facilities that contain food preparation areas (e.g., kitchens/cafeterias) have their own set of factors that can create indoor air quality problems.

Indoor air contaminants can originate from inside the building, and also from outside sources that are drawn inside. The EPA lists the factors most involved in the development of indoor air quality problems. These factors include:

- the source of odors or contaminants
- problems with the design or operation of the HVAC system

- the pathway between the source and the location of the complaint; and
- the building occupants

The sources from outside the building that can cause indoor air quality problems include contaminated outdoor air, emissions from nearby sources, soil gas and moisture or standing water.

- **Contaminated outdoor air** includes pollen, dust and fungal spores; industrial pollutants; and general vehicle exhaust.

- **Emissions from nearby sources**: vehicle exhaust; loading docks, dumpster odors and re-entrained—(i.e., air drawn back into the building), exhaust from the building itself or from neighboring buildings; and unsanitary debris near the outdoor air intake.

- **Soil gas** including radon; underground fuel tank leakage; previous use contaminants—landfills—and pesticides.

- **Moisture or standing water promoting excess microbial growth** which can be found in crawlspaces and rooftops after a rainfall.

Some indoor air quality problems can be solved quickly by in-house personnel, while others are complex and may require the assistance of professionals and/or mechanical engineers.


## EQUIPMENT

The equipment that is responsible for indoor air quality includes the HVAC system and non-HVAC equipment.


### HVAC System

Pollutant causing elements in the HVAC system include the dust or dirt in ductwork or other components; microbiological growth in drip pans, humidifiers, ductwork, coils; improper use of biocides, sealants, and/or cleaning compounds; improper venting of combustion products and refrigerant leakage.

**Non-HVAC Equipment**

The contaminants that can be found in non-HVAC equipment include emissions from office equipment (volatile organic compounds, ozone); supplies like solvents, toners, ammonia, etc., emissions from shops, labs and cleaning process; and elevator motors and other mechanical systems.

## HUMAN ACTIVITIES

The human activities that cause indoor air contaminants include: personal activities, housekeeping and maintenance activities.

- **Personal activities include** smoking, cooking, body odor and cosmetic odors.

- **Housekeeping activities include** cleaning materials and procedures; emissions from stored supplies or trash; use of deodorizers and fragrances; and airborne dust or dirt that is circulated by sweeping and vacuuming.

- **Maintenance activities include** microorganisms in mist from improperly maintained cooling towers; airborne dust or dirt; volatile organic compounds from use of paint, caulk, adhesives, and other products; pesticides from pest control activities and emissions from stored supplies.

## BUILDING COMPONENTS AND FURNISHINGS

Those items in a building that contribute to poor air quality include:

- **Locations that produce or collect dust or fibers** including textured surfaces such as carpeting, curtains and other textiles; open shelving; old or deteriorated furnishings; and materials containing damaged asbestos.

- **Unsanitary conditions and water damage** are conducted through microbiological growth on or in soiled water-damaged furnishings; microbiological growth in areas of surface condensation; standing water from clogged or poorly designed drains; and dry traps that allow the passage of sewer gas.

- **Chemicals released from building components or furnishings** include volatile organic compounds or inorganic compounds.

## OTHER SOURCES

Other sources of poor indoor air quality come from:

- **Accidental events** such as spills of water or other liquids; microbiological growth due to flooding or to leaks from roofs, and piping; and fire damage that includes soot, PCBs from electrical equipment and odors.

- **Special use areas and mixed use buildings** that contain smoking lounges; laboratories; print shops and art rooms; exercise rooms; beauty salons; and food preparation areas.

- **Redecorating/remodeling/repair activities** have emissions from new furnishings; dust and fibers from demolitions; odors and volatile organic and inorganic compounds from paint, caulk and adhesives; and microbiological releases from demolition or remodeling activities.

Small concentrations of these contaminants do not pose a risk for occupational exposure to occupants/tenants. Many of these contaminants are far below the standards or guidelines for occupational exposure.

### Common Indoor Pollutants

The following lists information about several indoor air pollutants common to offices, multi-use buildings and schools. The pollutants are discussed in terms of their effects on a company's employees and/or occupants and the control measures necessary for mitigation.

- Biological contaminants (mold, dust mites, pet dander, pollen, etc.)

- Carbon dioxide
- Carbon monoxide
- Dust
- Lead
- Nitrogen oxides
- Other volatile organic compounds (formaldehyde, solvents, cleaning agents)
- Pesticides
- Radon
- Tobacco smoke

### Figure 4-2. Common Indoor Air Pollutants

**BIOLOGICAL CONTAMINANTS**

**Description**
Common biological contaminants include: mold, dust mites, pet dander (skin flakes), droppings/body parts from roaches, rodents and other pests or insects, viruses, and bacteria. Many of these contaminants are small enough to be inhaled.

**Sources**
Biological contaminants are, or are produced by, living things. They are often found in areas that provide food and moisture or water. For example, damp or wet areas such as cooling coils, humidifiers, condensation pans, or unvented bathrooms can be moldy. Draperies, bedding, carpet, and other areas where dust collects may accumulate biological contaminants.

**Standards/Guidelines**
No federal government standards exist for biological contaminants in the indoor air environments for schools and office buildings as of 1999.

*(Continued)*

## Figure 4-2. Common Indoor Air Pollutants (Cont'd)

### Health Effects

Mold, dust mites, pet dander/pest droppings or body parts can trigger asthma. Biological contaminants, including molds and pollens can cause allergic reactions for many people. And, in addition, airborne diseases include tuberculosis, measles, staphylococcus infections, legionnaire's disease and influenza.

### Control Measures

Good housekeeping/maintenance of heating/air conditioning equipment is important, as is adequate ventilation and good air distribution. The key to mold control is moisture control. If mold is a problem, clean up the mold and get rid of excess water or moisture. Maintaining the relative humidity between 30%-60% helps control mold, dust mites, and roaches. Employ integrated pest management to control insect and animal allergens.

### CARBON DIOXIDE

### Description

Carbon dioxide ($CO_2$) is a colorless, odorless product of carbon combustion.

### Sources

Human metabolic processes and all combustion processes of carbon fuels are sources of $CO_2$.

### Standards/Guidelines

ASHRAE Standard 62-1989 recommends 1000 ppm as the upper limit for occupied classrooms/offices.

### Health Effects

$CO_2$ is an asphyxiate. At concentrations above 1.5% (15,000 ppm) some loss of mental acuity has been noted. (The recommended ASHRAE standard is 1000 ppm.)

## Figure 4-2. Common Indoor Air Pollutants (Cont'd)

### Control Measures

Ventilation with sufficient outdoor air controls $CO_2$ levels.

### CARBON MONOXIDE

### Description

Carbon monoxide (CO) is a colorless and odorless gas. It results from incomplete oxidation of carbon in combustion processes.

### Sources

Common sources of CO are from improperly vented furnaces, malfunctioning gas ranges, or exhaust fumes that have been drawn back into the building. Worn or poorly adjusted and maintained combustion devices (e.g. HVAC) can be significant sources, or a flue that is improperly sized, blocked, disconnected, or leaking. Auto, truck, or bus exhaust from attached garages, nearby roads, or idling vehicles in parking areas can also be a source.

### Standards or Guidelines

The OSHA standard for workers is 50 ppm for 1 hour. The US National Ambient Air Quality Standards for CO are 9 ppm for 8 hours and 35 ppm for 1 hour. The Consumer Product Safety Commission recommends levels not to exceed 15 ppm for 1 hour or 25 ppm for 8 hours.

### Health Effects

CO is an asphyxiate. Combined with hemoglobin, CO forms carboxyhemoglobin (COHb) which disrupts oxygen to tissues. Myocardium, brain, and exercising muscle tissues, those with the highest oxygen needs, are the first affected by COHb. The symptoms produced resemble influenza. These symptoms include fatigue, headache, dizziness, nausea and vomiting, cognitive impairment, and tachycardia. At high concentrations CO exposure can be FATAL.

## Figure 4-2. Common Indoor Air Pollutants (Cont'd)

### Control Measures

Combustion equipment must be maintained to assure that there are no blockages. Air and fuel mixtures must be properly adjusted to ensure more complete combustion. Vehicles adjacent to buildings should not be idling unnecessarily. Additional ventilation can be used as a temporary measure when high levels of CO are expected for short periods of time.

### DUST

### Description

Dust is made up of particles in the air that settle on surfaces. Large particles settle quickly and can be eliminated or greatly reduced by the body's natural defense mechanisms. Small particles are more like to be airborne and are capable of passing through the body's defense and entering the lungs.

### Sources

Many sources can produce dust including: soil, fleecy surfaces, pollen, lead-based paint, and burning wood, oil or coal.

### Standards or Guidelines

The EPA Ambient Air Quality standard for particles less than 10 microns is an annual average of 50 ng/m3 per hour and a 24-hour average of 150 ng/m3.

### Health Effects

Health effects vary depending upon the characteristics of the dust and any associated toxic materials. Dust particles may contain lead, pesticide residues, radon, or other toxic materials. Other particles may be irritants or carcinogens (e.g., asbestos).

## Figure 4-2. Common Indoor Air Pollutants (Cont'd)

**Control Measures**

Keep dust to a minimum. Use damp dusting and high efficiency vacuums. Upgrade filters in ventilation systems to medium efficiency; change frequently. Exhaust combustion appliance to the outside. Clean/maintain fuels, chimneys. Separate work areas from occupied ones during construction/remodeling.

### ENVIRONMENTAL TOBACCO SMOKE (ETS) OR SECONDHAND SMOKE

**Description**

Tobacco smoke consists of solid particles, liquid droplets, vapors and gases resulting from tobacco combustion. Over 4000 specific chemicals have been identified in the particulate and associated gases.

**Sources**

Tobacco product combustion.

**Standards or Guidelines**

Many office buildings/areas of public assembly have banned smoking indoors, or require specially designated smoking areas with dedicated ventilation systems. The Pro-Children Act of 1994 prohibits smoking in Head Start facilities, and in kindergarten, elementary and secondary schools that receive federal funding from the Department of Education the Department of Agriculture, or in the Department of Health and Human Services (except Medicare or Medicaid).

**Health Effects**

The effects of tobacco smoke on smokers include rhinitis/pharyngitis, nasal congestion, persistent cough, conjunctiva irritation, headache, wheezing, and exacerbation of

## Figure 4-2. Common Indoor Air Pollutants (Cont'd)

chronic respiratory conditions. The EPA classifies second-hand smoke as a "Group A" carcinogen. It has multiple health effects on children. It is also associated with the onset of asthma, increased severity of, or difficulty in controlling, asthma, frequent upper respiratory infections, persistent middle-ear effusion, snoring, repeated pneumonia, bronchitis.

### Control Measures

Smoke outside. Smoke only in rooms which are properly ventilated and exhausted to the outdoors.

### LEAD

### Description

Lead is a highly toxic metal.

### Sources

Sources of lead include drinking water, food, contaminated oil, dust, and air. Lead-based paint is a common source of lead dust.

### Standards or Guidelines

The Consumer Product Safety Commission has banned lead in paint.

### Health Effects

Lead can cause serious damage to the brain, kidneys, nervous system, and red blood cells. Children are particularly vulnerable. Lead Exposure in children can result in delays in physical development, lower IQ levels, shortened attention spans, and increased behavioral problems.

### Control Measures

Preventive measures to reduce lead exposure include: Cleaning, mopping floors, wiping window ledges and other smooth flat areas with damp cloths frequently.

## Figure 4-2. Common Indoor Air Pollutants (Cont'd)

### NITROGEN OXIDES

**Description**

The two most prevalent oxides of nitrogen are Nitrogen dioxide ($NO_2$) and nitric oxide (NO). Both are toxic gases with $NO_2$ being a highly reactive oxidant and corrosive.

**Sources**

The primary sources indoors are combustion processes, such as unvented combustion appliances, e.g., gas stoves, vented appliance with defective installations, welding, and tobacco smoke.

**Standards or Guidelines**

No standards have been agreed upon for the nitrogen oxides in indoor air. ASHRAE and the U.S.EPA National Ambient Air Quality Standards list 0.053ppm as the average 24-hour limit for $NO_2$ in outdoor air.

**Health Effects**

$NO_2$ is an irritant affecting the mucus of the eyes, nose, throat and respiratory tract. Extremely high-dose exposure (building fire) may result in pulmonary edema and diffuse lung injury. Continued exposure to high $NO_2$ levels contributes to acute or chronic bronchitis. Low level $NO_2$ exposure may cause increased bronchial reactivity in some asthmatics, decreased lung function in persons with chronic obstructive pulmonary disease and increased risk of respiratory infections.

**Control Measures**

Venting the $NO_2$ sources to the outdoors, and assuring that combustion appliances are correctly installed, used, and maintained are the most effective measures to reduce exposures.

## Figure 4-2. Common Indoor Air Pollutants (Cont'd)

**PESTICIDES**

**Description**
Pesticides are classed as semi-volatile organic compounds and include a variety of chemical in various forms. Pesticides are chemicals that are used to kill or control pests which include bacteria, fungi, and other organisms, in addition to insects and rodents. Pesticides are inherently toxic.

**Sources**
Pesticides occur indoors or can be tracked in from the outdoors.

**Standards or Guidelines**
No air concentration standards for pesticides have been set. However, EPA recommends Integrated Pest Management, which minimizes the use of chemical pesticides. Pesticide products must be used according to application and ventilation instructions provided by the manufacturer.

**Health Effects**
Symptoms may include headache, dizziness, muscular weakness, and nausea. Chronic exposure to some pesticides can result in damage to the liver, kidneys, endocrine and nervous systems.

**Control Measures**
Use Integrated Pest Management. If chemicals must be used, use only the recommended amounts, mix or dilute pesticides outdoors in an isolated well ventilated area, apply to unoccupied areas, and dispose of unwanted pesticides safely to minimize exposure.

## Figure 4-2. Common Indoor Air Pollutants (Cont'd)

**RADON**

**Description**

Radon is a colorless and odorless radioactive gas, the first decay product of radium-226. It decays into solid alpha particles which can be both inhaled directly or attached to dust particles that are inhaled. The unit of measure for radon is picocuries per liter (pCiL).

**Sources**

Radon exists in the earth's crust in widely varying concentrations. High concentrations of radon can occur in well water and in masonry blocks. The principle source is the earth around and under buildings. Radon penetrates cracks and drain openings in foundations, into basements and crawl spaces. Water containing radon will out-gas into spaces when drawn for use indoors. Some building materials will out-gas radon.

**Standards or Guidelines**

EPA recommends taking corrective action to mitigate radon if levels are at or exceed 4pCi/L.

**Health Effects**

Radon is a known human lung carcinogen. There is evidence of a synergistic effect between cigarette smoking and radon; the risks from exposure to both may exceed the risk from either acting alone.

**Control Measures**

Active Soil Depressurization and building ventilation are the two most commonly used strategies for controlling radon in buildings/schools. Sealing foundations to prevent radon entry as a stand-alone strategy is rarely successful. However sealing major entry points can improve the effectiveness of other strategies. Increased outdoor air ventila-

## Figure 4-2. Common Indoor Air Pollutants (Cont'd)

tion can reduce radon levels by dilution or pressurization of the building. A ventilation based strategy may not be the most effective strategy if initial radon levels are greater than 10pCi/L.

## VOLATILE ORGANIC CHEMICALS (FORMALDEHYDE, SOLVENTS, CLEANING AGENTS)

### Description

VOCs are emitted as gases from certain solids or liquids. VOCs include a variety of chemicals, some of which may have short- and long-term adverse health effects. Concentrations of many VOCs are consistently higher indoors (up to ten times higher) than outdoors.

### Sources

VOCs are emitted by a wide array of products numbering in the thousands. Examples include: paints, lacquers, paint strippers, cleaning supplies, pesticides, building materials, furnishings; office equipment such as copiers, printers, correction fluids, carbonless copy paper, graphics; craft materials, glues and adhesives, permanent markers, and photographic solutions.

### Standards or Guidelines

No standards have been set for VOCs in non-industrial settings. OSHA regulates formaldehyde as a carcinogen. OSHA adopted a Permissible Exposure Level (PEL) of .75ppm, and an action level of 0.5ppm. Formaldehyde should be mitigated when it is present at levels higher than 0.1ppm.

### Health Effects

Key signs or symptoms associated with exposure to VOCs include conjunctival irritation, nose and throat discomfort,

**Figure 4-2. Common Indoor Air Pollutants (Conclusion)**

headache, allergic skin reaction, dyspnea, declines in serum cholinesterase levels, nausea, emesis, epistaxis, fatigue, dizziness.

**Control Measures**
Increase ventilation when using products that emit VOCs. Meet or exceed any label precautions. Unopened containers of unused paints/similar materials should not be stored within the building. Formaldehyde can be readily measured. Identify and remove the source. If unable to remove, use a sealant on all exposed surfaces of paneling/other furnishings. An integrated pest management technique reduces the need for pesticides.

**Source**: United States Environmental Protection Agency, Indoor Air Quality, *Tools for Schools*. EPA 402-K-95-001 (Second Edition), August, 2000.

**Factors that Affect Occupant Comfort**
In addition to the factors that directly impact the levels of pollutants to which people are exposed, a number of environmental and personal factors can affect how people receive air quality. Some of these factors affect both the levels of pollutants and individual perception of air quality. According to the EPA, these factors include:

* Odors
* Temperature—too hot or cold
* Air velocity and movement—too drafty or stuffy
* Heat or glare from sunlight
* Glare from ceiling lights, especially on monitor screens
* Furniture crowding

- Stress in the workplace or home

- Feelings about physical aspects of the workplace: location, work environment, availability of natural light, and the aesthetics of office design, such as color and style

- Work space ergonomics, including height and location of computer, and adjustability of keyboards and desk chairs

- Noise and vibration levels

- Selection, location, and use of office equipment

**Productivity in the Workplace**

The environment of a building/facility workplace may have an effect on the productivity/morale of the occupant/tenant. A healthy environment, one that is adequate in terms of workspace that is ergonomically correct and appropriately sized for the designated task/job function may have a positive effect on the employee or occupant/tenant. Information technology tools along with good communication may also be contributing factors to the productivity/morale of the employee or occupant tenant.

In a study cited by the General Services Agency it was found that computer programmers, for example, performed optimally when the work area is efficiently designed. Access to windows, closed office space, workplace size, furniture and finishes were important or very important to the productivity of the occupant. (See Figure 4-3)

Thermal comfort, air quality and lighting may also have an impact on work performance. In studies cited by the GSA:

- The top quarter of computer programmers in a yearly coding competition "performed 2.6 times better in larger workspaces with fewer acoustic and visual disruptions than the bottom quarter, which worked in smaller spaces with less visual and acoustic control."

- Typewriting efficiency increased at 68 degrees than at 75 degrees F.

- The combination of a new building/facility along with individually controlled workstations saw a 16% increase in productivity whereas "disabling the workstation controls resulted in a 1.5% drop in productivity."

- Moving from a building with operable windows into a building with sealed windows saw increased absenteeism and decreased satisfaction.

Also important to tenant/occupant/worker satisfaction/morale is the organization itself—the workplace practices, management practices and style that is inherent to the organizational style.

## EFFECTIVE COMMUNICATION

The key to effective management is effective communication. It is also true with managing the IAQ of a facility. According to the EPA, the keys to effective communication with building occupants include:

### ENVIRONMENTAL FACTORS AND WORKER PERFORMANCE

| | Computer Programmers | |
| | Best Performers | Worst Performers |
|---|---|---|
| | *Top Quarter* | *Bottom Quarter* |
| Workstation size | *78 square feet* | *46 square feet* |
| Noise level acceptable | 57% | 29% |
| Privacy level acceptable | 62% | 19% |
| Phone can be silenced | 52% | 10% |
| Phone calls can be diverted | 76% | 19% |
| Frequent needless interruptions | 38% | 76% |

Courtesy: U.S. General Services Administration

**Figure 4.3**

- providing accurate information about factors that affect indoor air quality;

- clarifying the responsibilities of each party (e.g., building management, staff, tenants, contractors); and

- establishing an effective system for logging and responding to complaints should they occur.

**Provide Accurate Information**

Communication can prevent many indoor air quality problems if staff and building occupants understand how their activities affect indoor air quality. As noted earlier, a health and safety committee and/or a joint management-tenant IAQ task force may be set up to promote good working conditions. The purpose of the committee/task force is three-fold: It can help communicate information about indoor air quality; make the facility staff and management aware of current and/or potential air quality problems; and foster a sense of shared responsibility for maintaining a safe and comfortable indoor environment.

An air quality committee/task force will be successful if the following groups are included:

- building owner
- building manger
- facility personnel
- health and safety officials
- tenants and/or other occupants who are not facility staff
- representatives from the workforce (including union representatives)

**Clarify Responsibilities**

Of primary consideration is the inclusion into the employee handbook of the following: defining the responsibilities of building management, staff, and occupants and incorporating these responsibilities into the employee handbook and/or lease agreements. Responsibilities defined include:

- Use of space
- Occupancy Rate
- Modifications
- Notification of Planned Activities

## Use of Space

Permitted uses for specific areas within the facility and the maximum occupancy allowed should be communicated to the occupants. These permitted uses should be conveyed/communicated whenever there is a tenant change/move. As stated earlier, mixed use buildings pose the greatest venue for indoor air quality complaints. Kitchens and cooking odors, for example, may prove unpleasant to a nearby office occupant.

## Occupancy Rate

Ventilation systems are designed and operated to ensure air quality and comfort for a specific number of occupants. Therefore occupants/tenants should inform building management when there is an anticipated change in the number of occupants for a specific area. Occupants should be informed that the ventilation system in the building/facility is designed and operated for "x" number of occupants per ASHRAE Standard 62-1989 (Figure 4-4). This standard is the reference for proper ventilation required to provide a quality work environment.

## Modifications

Any change in the use of building space, including the installation of movable walls, partitions, new equipment etc., affect the ventilation requirements plans. Therefore these changes should be reviewed so that the HVAC system can be modified as needed. Building owners, facility mangers, and occupants share responsibility for monitoring new equipment installation and changes in the use of space.

## Notification of Planned Activities

A tenant/occupant notification procedure should be determined and established prior to beginning any activities such as

| Application | | Occupancy (people/1000ft²) | Cfm/ person | Cfm/ft² |
|---|---|---|---|---|
| **Food and Beverage Service** | Dining rooms | 70 | 20 | |
| | Cafeteria, fast food | 100 | 20 | |
| | Bars, cocktail lounges | 100 | 30 | |
| | Kitchen (cooking) | 20 | 15 | |
| **Offices** | Office space | 7 | 20 | |
| | Reception areas | 60 | 15 | |
| | Conference rooms | 50 | 20 | |
| **Public Spaces** | Smoking lounge | 70 | 60 | |
| | Elevators | | | 1.00 |
| **Retail Stores, Sales Floors, Showroom Floors** | Basement and street | 30 | | 0.30 |
| | Upper floors | 20 | | 0.20 |
| | Malls and arcades | 20 | | 0.20 |
| | Smoking lounge | 70 | 60 | |
| **Sports and Amusement** | Spectator areas | 150 | 15 | |
| | Game rooms | 70 | 25 | |
| | Playing floors | 30 | 20 | |
| | Ballrooms and discos 100 | 25 | | |
| **Theaters** | Lobbies | 150 | 20 | |
| | Auditoriums | 150 | 15 | |
| **Education** | Classroom | 50 | 15 | |
| | Music rooms | 50 | 15 | |
| | Libraries | 20 | 15 | |
| | Auditoriums | 150 | 15 | |
| **Hotels, Motels, Resorts, Dormitories** | Bedrooms | | | 30 cfm/room |
| | Living rooms | | | 30 cfm/room |
| | Lobbies | 30 | 15 | |
| | Conference rooms | 50 | 20 | |
| | Assembly rooms | 120 | 15 | |

Source: ASHRAE Standard 62-1989, Ventilation for Acceptable Air Quality, U.S. Environmental Protection Agency, Building Air Quality-A Guide for Facility Managers, 1991

**Figure 4-4. Selected Ventilation Recommendations ASHRAE Standard 62-1989**

painting, redecorating, controlling pests, remodeling, etc. Because these are activities that produce odors, the notification procedures should be established for informing tenants before the start of such activities.

**Response to Complaints**

Air quality is important to building tenants/occupants and owners. Therefore, any complaints by building tenants/occupants should be handled promptly and efficiently. As discussed earlier, procedures should be established for notification of complaints. Building occupants/tenants should know where to obtain complaint forms and how to notify the appropriate staff member(s) responsible for indoor air quality. In addition to the complaint form, establishing a record keeping system of complaints and actions taken is invaluable documentation that can help resolve complaints. This record keeping system also shows actual problem areas. By collecting this information in the facility manager/ IAQ manager has a "snapshot" of actual problem areas and the times these problems occur. The **IAQ Complaint Form and Incident Log** (Figure 4-5) that follows (and also reproduced in the Appendix) can be used to track complaints related to the indoor environment.

## COMMUNICATION—
## A KEY TO RESOLVING IAQ PROBLEMS

Indoor air quality problems can sometimes be identified and resolved quickly. On other occasions, complaints originate from the interaction of several variables, and detailed investigation may be necessary in order to resolve the problem. Whatever the case, complaints should be responded to quickly, respecting the view of the complainant.

**The Importance of Response**

IAQ complaints may be grounded in poor indoor air quality, thermal conditions, noise, glare, or even job stresses. However, to

**Sample Form**
**Indoor Air Quality Complaint Form**

This form should be used if the complaint may be related to indoor air quality. Indoor air quality problems include concerns with temperature control, ventilation, and air pollutants. Personal observations can help to resolve the problem as quickly as possible. Please use the space below to describe the nature of the complaint and any potential causes.




We may need to contact you to discuss the complaint. What is the best time to reach you?

_______________________________________________________

So that we can respond promptly, please return this form to:

_______________________________________________________

IAQ Manager or Contact Person

**Figure 4-5.** *(Courtesy: Environmental Protection Agency)*

establish open communication with building occupants/tenants, it is in the best interest of the building owner/manager to listen to and respond to each complaint. Open communication can alleviate the perception that no action is being taken or that important information is being withheld.

Being attentive to communication and problem solving helps to ensure the support and cooperation of building occupants as the complaint is investigated and resolved. The message that management believes that a good indoor air quality is an essential component of a healthy indoor environment ensures the support and cooperation of the building occupants/tenants as the complaint is investigated and resolved.

According to the EPA, communications, whether they occur in conversations or in writing, should include the following information:

- types of complaints received
- management's environment policy
- management's response to occupant complaints

- management's investigation and mitigation plans (including hiring outside consultants)

- the names and telephone numbers of appropriate staff responsible for investigating and mitigating employee/occupant complaints/questions. This staff may include members of the facility staff, medical, and/or health and safety staff.

**Maintaining the Lines of Communications**

Proper identification of the IAQ contact personnel needs to be communicated to the occupant/tenants. This can be done in a variety of ways: building postings, newsletters, directory information, etc. Complaints may be channeled through a committee, supervisor, health and safety representative, or directly to the IAQ representative. IAQ complaints, as discussed earlier, should be handled quickly and responsibly. Minor complaints such as annoying odors from an easily-identified source can be handled promptly. More serious complaints/concerns may take longer to resolve. This information must be communicated to the occupants, as well as to the complainants.

Figure 4-6 is a sampling of the types of IAQ problem and their associated responses. It is a sampling of the type of problems that require immediate action and those that are less problematic. All complaints/problems should be responded to no matter what the degree of complaint is rated—either high priority or low priority.

**SELECTED INDOOR AIR QUALITY PROBLEMS**

| | | |
|---|---|---|
| **Problems Requiring Immediate Action** | **Complaints of headaches, nausea, and combustion odors** | Carbon monoxide poisoning is a possibility. Immediately investigate sources of combustion gases. |
| | **One or more occupants of the building have been diagnosed with Legionnaire's disease** | This is a potentially life-threatening illness. Request Health Department assistance in determining whether the building is the source of the infection. |
| | **Water from a roof leak has flooded a portion of the carpeting.** | Discard any damp carpeting that cannot be lifted, and dried within a short period of time. Use proper cleaning/disinfecting procedures to prevent mold/mildew growth that could cause serious IAQ problems. |
| **Problems That Require A Response, But Are Not Emergencies** | **Inspection of the humidification system reveals an accumulation of slime & mold. No reported health complaints to suggest IAQ problems.** | Inadequately maintained humidifiers can promote the growth of biological contaminants. Equipment should be cleaned thoroughly and modification of maintenance practices should be considered. |

| | |
|---|---|
| **A group of occupants share common symptoms of headaches, eye irritation, and respiratory complaints. They decided that their problems are due to conditions in the building.** | These symptoms suggest an IAQ problem that is not life-threatening. A prompt response to thiscomplaint is warranted. |
| **Immediately after delivery of new furnishings (furniture or carpeting), occupants complain of odors and discomfort.** | Volatile compounds emitted by the new furnishings could be causing the complaints. |
| **Local news articles suggest that some buildings in the area have high indoor radon levels.** | Test appropriate locations in the building to determine the indoor radon concentration. |
| **Consider if some old pipe insulation contains asbestos.** | Contact a laboratory that tests asbestos. |

Source: United States Environmental Protection Agency.

**Figure 4-6. Selected Indoor Air Quality Problems**

Potentially serious problems and/or widespread contamination need the attention of the health and safety committee and/or a joint management-tenant IAQ task force. During the investigation, information should be conveyed to the occupants/tenants to avoid confusion and hard feelings. Representatives from the occupants/tenants should also be part of the investigation/mitigation procedure.

## Improving Channels of Communication

Basic information about the IAQ problem and mitigation efforts should be given to building occupants/tenants to avoid fears, rumors and suspicions. Factual information during the investigation in the form of notices, memos and/or newsletter articles should keep tenants/occupants up-to-date, informed, cooperative and less suspicious about the investigative activities. Efforts should be made to give the occupants/tenants only the information needed. Releasing premature information, i.e., information that is incomplete or an incorrect representation of risk can be detrimental. The facility manager should clear all communications with the building owner/management, and/or legal counsel.

When disseminating information, management should be factual and to the point using the following guidelines suggested by the EPA:

- The definition of the complaint.
- The progress of the investigation.
- Factors that have been evaluated.
- How long the investigation might take.
- Attempts that are being made to improve indoor air quality.
- Work that remains to be done and the schedule for its completion.

*The Definition of the Complaint*

Define the complaint area based upon the location and distribution of complaints. This information may be revised as the investigation progresses.

*The Progress of the Investigation*

Include in this information the types of information that is being gathered. Elicit support and commitment from the occupants/tenants in this area of the investigation.

*Evaluation Factors*

List the factors that have been determined to be the causes and/or contributing to the problem, as well as those factors that have been eliminated as contributing to, or actually causing the problem.

*Length of the Investigation*

This section should include the estimated time for discovery and mitigation of the problem.

*Attempts Being Made*

Occupants/tenants should know what attempts are being made to improve indoor air quality.

*Schedule for Completion*

Identify projected or anticipated completion date for problem resolutions.

In summary, communication is essential for an effective IAQ management policy. Collecting the information, responding quickly to both minor and major complaints and disseminating the right information at the proper time, leads to a healthy relationship between management and occupants/tenants.

**The IAQ Profile**

An indoor air quality profile is an effective tool to aid in the understanding of the current status of air quality in the building and provides the necessary information of those factors. It can help assess the status of the current air quality, as well as help in identifying potential problem areas. It can also help the budgeting function for current and future maintenance and/or modifications. Combining the IAQ profile with the security function, lighting management systems, etc., provides a workable picture or "blueprint" of the major building systems.

Key elements of an IAQ profile include:

- Original function of the building.

- Current function of the building.

- Changes that have occurred.

- Changes that are needed.

*Original Function of the Building*

Research the original use of the building including its components and furnishings, mechanical equipment (HVAC and non-HVAC), and the occupant population and associated activities.

*Current Function of the Building*

If the building was commissioned compare the information from the commissioning to its current use.

*Changes that Have Occurred*

Analyze past and present building layout changes. Include the HVAC system in the analysis and determine if the HVAC system reflects the current usage.

*Changes that Are Needed*

Determine the changes that may be needed to prevent future IAQ problems. Potential changes in future uses should be considered.

## CREATING AN IAQ PROFILE

The following basic list should be used by the staff person or persons who can understand the necessary data and who will be responsible for collecting the needed information/data:

- Basic understanding of HVAC system operating principles.

- Ability to read plans both architectural and mechanical.

- Collect the manufacturer's catalog equipment data.

- Identify items of office equipment.

- Cooperate with building occupants to gather information about space usage.

- Collect information about the HVAC system operation.

- Collect information about equipment condition and maintenance schedules.

- Collect information from subcontractors including outsourced cleaning and pest control companies about their schedules and materials used.

- Collect and understand the information in the Material Safety Data Sheets (MSDSs).

To ensure adequate indoor air quality for the building/facility in their charge, facility managers may wish to develop a building air quality action plan to aid in the assessment and mitigation of their building/facility's existing air quality.

A sample Building Air Quality Action Plan Verification Checklist developed by the Environmental Protection Agency (EPA) and the National Institute of Occupational Safety and Health (NIOSH) appears in Appendix III.


## SAMPLE IAQ PROBLEMS AND SOLUTIONS

Some of the common problems associated with commercial buildings/facilities follow. (See Figure 4-8). These examples are just that—problems that could or could not be associated with any one particular building. They are listed here as examples of problems along with their solutions.

| File # | Date | Location | Complaint | Interview | Activity Log | Room Record | HVAC Checklist | Pollutant Pathway | Source | Hypotheses | Outcome | Entry by |
|---|---|---|---|---|---|---|---|---|---|---|---|---|
| | | | | | | | | | | | | |
| | | | | | | | | | | | | |
| | | | | | | | | | | | | |
| | | | | | | | | | | | | |
| | | | | | | | | | | | | |

Investigation Record (check the forms that were used)

**Figure 4-7. Sample Form Incident Log**

1   Outdoor air ventilation rate is too low.

2   Overall ventilation rate is high enough, but poorly distributed and not sufficient in some areas.

3   Contaminant enters building from outdoors.

4   Occupant activities contribute to air contaminants or to comfort problems.

5   HVAC system is a source of biological contaminants.

6   HVAC system distributes contaminants.

7   Non-HVAC equipment is a source or distribution mechanism for contaminants.

8   Surface contamination due to poor sanitation or accidents.

9   Mold and mildew growth due to moisture from condensation.

10   Building materials and furnishings produce contaminants.

11   Housekeeping or maintenance activities contribute to problems.

12   Specialized use areas as sources of contaminants. Remodeling or repair activities reproduce problems.

13   Remodeling or repair activities produce problems.

14   Combustion gases.

15   Serious building-related illness.

---

Courtesy:     U.S Environmental Protection Agency
Office of Air and Radiation.
Office of Atmospheric and Indoor Air Programs.
Indoor Air Division;
U.S. Department of Health and Human Services
Public Health Service
Centers for Disease Control
National Institute for Occupational Safety and Health. *Building Air Quality A Guide for Building Owners and Facility Managers,* 1991.

**Figure 4-8. Sample IAQ Problems and Solutions**

### Figure 4-8. Sample IAQ Problems and Solutions (*Continued*)

---

## SAMPLE IAQ SOLUTIONS

**Problem #1: Outdoor Air Ventilation Rate is Too Low**

**Examples:**

*Routine odors from occupants and normal office activities result in problems* such as drowsiness, headaches, discomfort.

*Measured outdoor air ventilation rates do not meet guidelines for outdoor air supply* that is design specifications, applicable codes, or ASHRAE 62-1989.

**Peak $CO_2$ concentrations above 1000 ppm indicate inadequate ventilation.**

**Corrosion of fan casing causes air bypassing and reduces airflow in system.**

**Solutions**

*Open, adjust or repair air distribution system*
- outdoor air intakes
- mixing and relief dampers
- supply diffusers
- fan casings

*Increase outdoor air within the design capacity of*
- air handler
- heating and air conditioning equipment
- distribution system

*Modify components of the HVAC system as needed to allow increased outdoor air*
    (e.g., increase capacity of heating and cooling coils)

*Design and install an updated ventilation system*

*Reduce the pollutant and/or thermal load on the HVAC system*

### Figure 4-8. Sample IAQ Problems and Solutions (*Continued*)

- reduce the occupant density: relocate some occupants to other spaces to redistribute the load on the ventilation system
- relocate or reduce usage of heat-generating equipment

### Problem #2: Overall Ventilation Rate is High Enough, But Poorly Distributed and Not Sufficient in Some Areas

### Examples
*Measured outdoor air meets guidelines at building air inlet, but there are zones where heat, routine odors from occupants, and normal office activities result in complaints*
(e.g., drowsiness, headaches, comfort complaints)

### Solutions
*Open, adjust, or repair air distribution system*
- supply diffuser
- return registers

*Ensure proper air distribution*
- balance the air handling system
- make sure that there is an air gap at tops and bottoms of partitions to prevent dead air space
- relocate supply and/or return diffusers to improve air distribution

*Seal leaky ductwork*

*Remove obstructions from return air plenum*

*Control pressure relationships*
- install local exhaust in problem areas and adjust HVAC system to provide adequate make-up air
- move occupants closer to supply diffusers
- relocate identified contaminant sources closer to exhaust intakes

## Figure 4-8. Sample IAQ Problems and Solutions (*Continued*)

*Reduce source by limiting activities or equipment use that
produce heat, odors or contaminants*
*Design and install an appropriate ventilation system*

**Problem #3: Outdoor Contaminants Enters Building**

**Examples**
*Soil gases*
(e.g., radon, gasoline from tanks, methane from landfills)
*Contaminants from nearby activities*
(e.g., roofing, dumpster, construction)
*Outdoor air intake near source*
(e.g., parking, loading dock, building exhaust)
*Outdoor air contains pollutants or excess moisture*
(e.g., cooling tower mist entrained in outdoor air intake)

**Solutions**
*Remove the source (if it can be moved easily)*
- remove debris around outdoor air intake
- relocate dumpster

*Reduce source (for example, shift time of activity to avoid
occupied periods)*
- painting, roofing, demolition
- housekeeping, pest control

*Relocate elements of the ventilation system that contribute
to entry of outdoor air contaminants*
- separate outdoor air intakes from sources of odors, contaminants
- separate exhaust fan outlets from operable windows,
doors, air intakes
- make rooftop exhaust outlets taller than intakes

## Figure 4-8. Sample IAQ Problems and Solutions (*Continued*)

*Change air pressure relationships to control pollutant pathways*

- install sub-slab depressurization to prevent entry of soil gas contaminant (radon, gases from landfills and underground tanks)
- pressurize the building interior relative to outdoors (this will not prevent contaminant entry at outdoor air intakes)
- close pollutant pathways (e.g., seal cracks and holes)

*Add special equipment to HVAC system*

- filtration equipment to remove pollutants (select to fit the situation)

*Add special equipment to HVAC system*

## Problem #4: Occupant Activities Contribute to Air Contaminants/Comfort Levels

### Examples

*Smoking*

*Special activities such as print shops, laboratories, kitchens*

*Interference with HVAC system operation:*

- blockage of supply diffusers to eliminate drafts
- turning off exhaust fans to eliminate noise
- use of space heaters, desktop humidifiers to remedy local discomfort. (*Note*: While such interference can cause IAQ problems, it is often initiated in response to unresolved ventilation or temperature control problems.)

### Solutions

*Remove the source by eliminating the activity*

(*Note*: This may require a combination of policy-setting and educational outreach.)

- smoking

## Figure 4-8. Sample IAQ Problems and Solutions (*Continued*)

- use of desktop humidifiers and other personal HVAC equipment
- unsupervised manipulation of HVAC system

*Reduce the source*
- select materials and processes which minimize release of contaminants while maintaining adequate safety and efficacy (e.g., solvents, art materials)

*Install new or improved local exhaust to accommodate the activity, adjust HVAC system to ensure adequate make-up air, and verify effectiveness*
- smoking lounge, storage areas which contain contaminant sources
- laboratory hoods, kitchen range hoods (venting to outdoors, not re-circulating.

### Problem #5: HVAC System is a
### Source of Biological Contaminants

The HVAC system can act as a source of contaminants by providing a hospitable environment for the growth of microorganisms and then distributing biologically-contaminated air within the building.

### Examples
*Surface contamination by molds (fungi), bacteria*
- drain pans
- interior of ductwork
- air filters and filter media (collected debris)

Solutions
*Remove source by improving maintenance procedures*
- inspect equipment for signs of corrosion, high humidity
- replace corroded parts
- clean drip pans, outdoor air intakes, other affected locations

## Figure 4-8. Sample IAQ Problems and Solutions (*Continued*)

- use biocides, disinfectants, and sanitizers with extreme caution and ensure that occupant exposure is minimized

*Provide access to all the items that must be cleaned, drained, or replaced periodically*

### Problem #6: HVAC System Distributes Contaminants

### Examples

*Unfiltered air bypasses filters due to problems*
- filter tracks are loose
- poorly maintained filters sag when they become overloaded with dirt
- filters are the wrong size

*Re-circulation of air that contains dust or other contaminants*
- system re-circulates air from rooms containing pollutant sources
- return air plenum draws air from rooms that should be exhausted (e.g., janitor's closets)
- return air plenums draw soil gases from interiors of block corridor walls that terminate above ceilings

### Solutions

*Modify air distribution system to minimize re-circulation of contaminants*
- provide local exhaust at point sources of contaminants, adjust HVAC system to provide adequate make-up air, and test to verify performance
- increase proportion of outdoor air
- seal unplanned openings into return air plenums and provide alternative local ventilation (adjust HVAC system to provide adequate make-up air and test to verify performance)

### Figure 4-8. Sample IAQ Problems and Solutions (*Continued*)

*Improve housekeeping, pest control, occupant activities, and equipment use to minimize release of contaminants from all sources*
*Install improved filtration equipment to remove contaminants*
*Check filter tracks for any gaps*

### Problem #7: Non-HVAC Equipment is a Source or Distribution Mechanism for Contaminants

These examples/solutions refer to medium- to large-scale pieces of equipment.

### Examples

Non-HVAC equipment can produce contaminants, as in the case of:

- wet process copiers
- large dry process copiers
- engineering drawing reproduction machines

*It can also distribute contaminants, as in the case of:*
- elevators, which can act as pistons and draw contaminants from one floor or another

### Solutions

*Install local exhaust near machines*
(*Note*: Adjust HVAC system to provide adequate make-up air, and test to verify performance.)

*Reschedule use to occur during periods of low occupancy*

*Remove source*
- relocate occupants out of rooms that contain contaminant-generating equipment
- relocate equipment into special use areas equipped with effective exhaust ventilation (test to verify control of air pressure relationships)

### Figure 4-8. Sample IAQ Problems and Solutions (*Continued*)

*Change air pressure relationships to prevent contaminants from entering elevator shaft*

**Problem #8: Surface Contamination Due to Poor Sanitation or Accidents**

**Examples**
Biological contaminants result in allergies or other diseases
- fungal, viral, bacterial (whole organisms or spore)
- bird, insect, or rodent parts or droppings, hair, dander (in HVAC, crawlspace, building shell, or near outdoor air intakes)

**Accidents**
- spills of water, beverages, cleansers, paints, varnishes, mastics or specialized products (printing, chemical art supplies)
- fire damage: soot, odors, chemicals

**Solutions**
*Clean*
- HVAC system components
- some materials and furnishings (others may have to be discarded) (*Note*: Use biocides, disinfectants, and sanitizers with caution and ensure that occupant exposure is minimized.)

*Remove sources of microbiological contamination*

- water-damaged carpet, furnishings, or building materials

*Modify environment to prevent recurrence of microbiological growth*
- improve HVAC system maintenance
- control humidity or surface temperatures to prevent condensation

## Figure 4-8. Sample IAQ Problems and Solutions (*Continued*)

*Provide access to all items that require periodic maintenance*

*Use local exhaust where corrosive materials are stored*

*Adjust HVAC system to provide adequate make-up air, and test to verify performance*

**Problem #9: Mold and Mildew Growth Due to Moisture from Condensation**

**Examples**
*Interior surfaces of walls near thermal bridges*
(e.g., un-insulated locations around structural members)

*Carpeting on cold floors*

*Locations where high surface humidity promotes condensation*

**Solutions**
*Clean and disinfect to remove mold and mildew*
(*Note*: Take actions to prevent recurrence of microbiological contamination. Biocides, disinfectants, and sanitizers should be used with caution. Minimize occupant exposure.)

*Increase surface temperatures to treat locations that are subject to condensation*
- insulate thermal bridges
- improve air distribution

*Reduce moisture levels in locations that are subject to condensation*
- repair leaks
- increase ventilation (in cases where outdoor air is cold and dry)
- dehumidify (in cases where outdoor air is warm and humid)

## Figure 4-8. Sample IAQ Problems and Solutions (*Continued*)

*Dry carpet or other textiles promptly after steam cleaning* (*Note*: Increase ventilation to accelerate drying.)

*Discard contaminated materials*

**Problem #10: Building Materials and Furnishings Produce Contaminants**

**Examples**
*Odors from newly installed carpets, furniture, wall coverings*
*Newly dry-cleaned drapes of other textiles*

**Solutions**
*Remove source with appropriate cleaning methods*
- steam clean carpeting and upholstery, then dry quickly, ventilating to accelerate the drying process
- accept only fully dried, odorless dry-cleaned products

*Encapsulate source*
- seal surface of building materials that emit formaldehyde

*Reduce source*
- schedule installation of carpet, furniture, and wall coverings to occur during periods when the building is unoccupied
- have supplier store new furnishings in a clean, dry, well-ventilated area until VOC outgassing has diminished

*Increase outdoor air ventilation*
- total air supplied
- proportion of fresh air

*Replace emission producing materials with lower emission alternatives*
(*Note*: Only limited information on emissions from materials is available at this time. Purchasers can request that suppliers

## Figure 4-8. Sample IAQ Problems and Solutions (*Continued*)

provide emissions test data, but should use caution in interpreting the test results.)

### Problem #11: Housekeeping or Maintenance Activities Contribute to Problems

### Examples

*Cleaning products emit chemicals, odors*
*Particulates become airborne during cleaning* (e.g., sweeping, vacuuming)
*Contaminants are released from painting, caulking, lubricating*
*Frequency of maintenance is insufficient to eliminate contaminants*

### Solutions

*Remove source by modifying standard procedures or frequency of maintenance*
(*Note*: Policy-setting and training in IAQ may be required whenever there is a change in policy.)

- improve storage practices
- change the time of painting, cleaning, pest control, and other contaminant-producing activities during unoccupied periods.
- make maintenance easier by improving access to filters, coils, and other components

### Reduce source

- select materials to minimize emissions of contaminants while maintaining adequate safety and efficacy
- use portable HEPA ("high efficiency particulate arrestance") vacuums vs. low-efficiency paper-bag collections

### Use local exhaust

- on a temporary basis to remove contaminants from work areas

### Figure 4-8. Sample IAQ Problems and Solutions (*Continued*)

- as a permanent installation where contaminants are stored.

**Problem #12: Specialized Use Areas as Sources of Contaminants**

**Examples**
*Food preparation*
*Art or print rooms*
*Laboratories*

**Solutions**
*Change pollutant pathway relationships*
- run specialized use area under negative pressure relative to surrounding areas
- install local exhaust, adjust HVAC system to provide make-up air, and test to verify performance

*Remove source by ceasing, relocating, or rescheduling incompatible activities*
*Reduce source by selecting materials to minimize emissions of contaminant while maintaining adequate safety and efficacy*
*Reduce source by using proper sealing and storage for materials that emit contaminants*

**Problem #13: Remodeling or Repair Activities Produce Problems**

**Examples**
*Temporary activities produce odors and contaminants*
- installation of new particleboard, partitions, carpet, or furnishings
- painting
- reroofing
- demolition

*Existing HVAC system does not provide adequate ventilation for new occupancy or arrangement of space*

### Figure 4-8. Sample IAQ Problems and Solutions (*Continued*)

**Solutions**

*Modify ventilation to prevent re-circulation of contaminants*
- install temporary local exhaust in work area, adjust HVAC system to provide make-up air, and test to verify performance
- seal off returns in work area
- close outdoor air damper during re-roofing

*Reduce source by scheduling work for unoccupied periods and keeping ventilation system in operation to remove odors and contaminants*

*Reduce source by careful materials selection and installation*
- materials with minimal contaminant emissions should be selected
- new furnishings should be stored in a clean, dry, well-ventilated area until VOC outgassing has diminished
- adhesives used in the installation procedures should have limited emissions of contaminants

*Modify HVAC or wall partition layout if necessary*
- partitions should not interrupt airflow
- relocate supply and return diffusers
- adjust supply and return air quantities
- adjust total air and/or outdoor air supply to serve new occupancy

**Problem #14: Combustion Gases**

Combustion odors can indicate the existence of a serious problem. One combustion product, carbon monoxide, is an odorless gas. Carbon monoxide poisoning can be life-threatening.

**Examples**

*Vehicle exhaust*
- offices above (or connected to) an underground parking garage

## Figure 4-8. Sample IAQ Problems and Solutions (*Continued*)

- rooms near (or connected by pathways to) a loading dock or service garage.

***Combustion gases from equipment***
(e.g., spillage from inadequately vented appliances, cracked heat exchanger, re-entrainment because local chimney is too low)

- areas near a mechanical room
- distributed throughout zone or entire building

**Solutions**

***Seal to remove pollutant pathway***

- close openings between the contaminant source and the occupied space
- install well-sealed doors with automatic closers between the contaminant source and the occupied space

***Remove source***

- improve maintenance of combustion equipment
- modify venting or HVAC system to prevent back drafting
- relocate holding area for vehicles at loading dock, parking area
- turn off engines of vehicles that are waiting to be unloaded

***Modify ventilation system***

- install local exhaust in underground parking garage (adjust HVAC system to provide make-up air and test to verify performance)
- relocate fresh air intake (move away from source of contaminants)
- elevate chimney exhaust outlet

***Modify pressure relationships***

- pressurize spaces around area containing source of combustion gases

## Figure 4-8. Sample IAQ Problems and Solutions (*Continued*)

**Problem #15: Serious Building-Related Illness**
Some building-related illnesses can be life-threatening. Even a single confirmed diagnosis (which involves results from specific medical tests) should provoke an immediate and vigorous response.

**Examples**
>*Legionnaire's disease*
>(*Note*: If Legionnaire's disease is suspected, call the local public health department, check for obvious problem sites, and take corrective action. This disease is not necessarily associated with building occupancy. Public health agencies do not investigate single cases, however, check for new cases.

>*Hypersensitivity pneumonitis*
>(*Note*: Affected occupant(s) should be removed and may not be able to return unless the causative agent is removed from the affected person's environment.)

**Solutions**
>*Work with public health authorities*
>- evacuation may be recommended or required.

>*Remove source*
>- drain, clean, and decontaminate drip pans, cooling towers, room unit air conditioners, humidifiers, dehumidifiers, and other habitats of *Legionella*, fungi, and other organisms using appropriate protective equipment.
>- install drip pans that drain properly
>- provide access to all the items that must be cleaned, drained, or replaced periodically
>- modify schedule and procedures for improved maintenance

>*Discontinue processes that deposit potentially contaminated moisture in air distribution system*

**Figure 4-8. Sample IAQ Problems and Solutions (*Concluded*)**

- air washing
- humidification
- cease nighttime shutdown of air handlers

**Sources:**

United States Environmental Protection Agency, *Indoor Air Quality, Tools for Schools*. EPA 402-K-95-001 (Second Edition), August, 2000.

United States Environmental Protection Agency, *Building Air Quality—A Guide for Facility Managers*. EPA 400/1-91/033 December, 1991.

United States Environmental Protection Agency, *Building Air Quality Action Plan*. EPA 402-K-98-001, June, 1998.

# Chapter 5
# *Emergency Preparedness*

## KEY ELEMENTS IN EMERGENCY PREPAREDNESS

The six key elements integral to any emergency response plan are:

1.  Emergency escape procedures and emergency route assignments.

2.  Procedures to be followed by employees who remain to perform (or shut down) critical plant operations before they evacuate.

3.  Procedures to account for all employees after emergency evacuations have been completed.

4.  Rescue and medical duties for those employees who are to perform them.

5.  The preferred means for reporting fires and other emergencies.

6.  Names or regular job titles of people or departments to be contacted for further information or explanation of duties under the plan.

Additionally, employers are required to establish and install an alarm system that provides warning for necessary action as called for in the emergency action plan, or provides reaction time to ensure the safe escape of employees/occupants from the workplace, the immediate work area, or both. If the alarm system is used for alerting fire brigade members, or for other purposes, separate and distinct signals must be used for each purpose.

Finally, before implementing the emergency action plan, the employer must designate and train a sufficient number of people to assist in the safe and orderly emergency evacuation of employees and occupants.

## PLANNING FOR POTENTIAL DISASTERS

Emergency preparedness planning should address all potential emergencies and occurrences that can be expected. Therefore, it is essential to perform a hazard audit and worksite or facility analysis to determine potential disasters. For example, information on chemicals that are used in the worksite can be obtained from the Material Safety Data Sheets (MSDS) that are provided by the manufacturer or supplier of the chemical. These forms describe the hazards that a chemical may present, list precautions to take when handling, sorting or using the substance, and outline emergency and first-aid procedures.

### Response Effectiveness and Leadership

The effectiveness of response during any emergency, including a disaster, depends on the amount of planning and training that occurs. Because management must show its active and aggressive support for emergency preparedness and disaster planning, it is management's responsibility to ensure that a program is established and that it is frequently reviewed and updated. The input and support of all personnel within the facility—including company employees and tenant-occupants—*must be obtained* to ensure an effective program. The emergency preparedness plan should be developed locally—i.e., it must be site-specific, and the plan should be comprehensive enough to address all types of emergencies.

Those employees who remain behind to care for essential facility functions until their evacuation becomes necessary must also be provided with specific procedures that detail what actions they must take. These detailed procedures may include monitoring critical power supplies, as well as other essential services that cannot be shut down for every emergency alarm.

In situations that call for emergency evacuation, floor plans and/or workplace maps that clearly show the emergency escape route and safe areas must also be included in the plan.

The emergency preparedness and disaster plan must be reviewed with all employees and tenant-occupants:

- When the plan is initially developed.
- Whenever new employees and/or occupants are assigned to the facility.
- Whenever responsibilities under the plan change.
- Whenever the plan itself is changed.

## The Chain of Command

A chain of command should be established to minimize any confusion among employees and tenant-occupants regarding the lines of decision-making authority. Responsible people should be selected to coordinate the work of an emergency response team. Facility managers, by virtue of their critical roles within an organization, are the logical people to be charged with this responsibility. However, it is imperative that additional personnel be designated and trained as back-up support so that the facility is covered, available to implement the plan at all times. The responsibilities of the designated person and back-up support include:

1. Assessing the situation to determine whether an emergency or disaster exists that requires the activation of emergency procedures.

2. Directing all site-evacuation efforts to minimize injuries and property loss.

3. Ensuring that outside emergency services (i.e., medical, fire and police) are notified when necessary.

4. Directing facility/plant operations shutdown, when necessary.

**Communications**

When an emergency or disaster is determined, it may become necessary to evacuate an entire facility, including offices. In situations involving a fire or explosion, normal utility services, e.g., electricity, gas, heating oil, and water, as well as telephone services may become non-existent. For this reason, an alternate "safe" site should be considered by companies. A "safe" site not only provides employers with a "central command station" that can be used to coordinate emergency response, but it also provides a centralized reporting location that can be used to expedite employee/personnel/occupant accounting.

Various communications equipment such as portable radio units and systems, public address systems, cellular telephones, auto dialers and other communications transmission systems should be located at the "safe" site for use in an emergency and/or disaster situation.

Alarm systems, as required by various standards and/or codes, must be installed so that employees and tenant-occupants can be alerted to the emergency and/or disaster threat. While the type of alarm system is contingent upon its purpose, alarms should be audible or be seen by all people in the facility. An auxiliary power supply is required in the event that the main power source is interrupted. An alarm should be both distinctive and recognizable as a signal to either evacuate the work area or facility, or to take other actions as prescribed in the emergency preparedness plan. Employers who must comply with an OSHA standard that requires emergency employee alarms should note the following OSHA requirements.

| Subject And<br>Standard Number | Emergency Employee Alarms:<br>OSHA Requirements |
|---|---|
| Employee<br>Alarm Systems<br>1819.165(a) | (a) Scope And Application. (1) This section applies to all emergency employee alarms installed to meet a particular OSHA stan- |

dard. This section does not apply to those discharge or supervisory alarms required on various fixed extinguishing systems or to supervisory alarms on fire suppression, alarm or detection systems unless they are intended to be employee alarm systems.

(2) The requirements in this section that pertain to maintenance, testing and inspection shall apply to all local fire alarm signaling systems used for alerting employees regardless of the other functions of the system.

(3) All pre-discharge employee alarms installed to meet a particular OSHA standard shall meet the requirements of paragraphs (b) (12) through (4), (c), and (d) (1) of this section.

1910.165(b)  (b) General Requirements. (1) The employee alarm system shall provide warning for necessary emergency action as called for in the emergency action plan, or for reaction time for safe escape of employees from the workplace or the immediate work area, or both.

(2) The employees' alarm shall be capable of being perceived above ambient noise or light levels by all employees in the affected portions of the workplace. Tactile devices may be used to alert those employees who would not otherwise be able to recognize the audible or visual alarm.

(3) The employee alarm shall be distinctive and recognizable as a signal to evacuate the work area or to perform actions designated under the emergency action plan.

(4) The employer shall explain to each employee the preferred means of reporting emergencies, such as manual pull box alarms, public address systems, radio or telephones. The employer shall post emergency

telephone numbers near telephones, or employee notice boards, and other conspicuous locations when telephones serve as the employee alarm system, all emergency messages shall have priority over all non-emergency messages.

(5) The employer shall establish procedures for sounding emergency alarms in the workplace. For those employers with 19 or fewer employees in a particular workplace, direct voice communication is an acceptable procedure for sounding the alarm provided all employees can hear the alarm. Such workplaces need not have a back-up system.

1910.165(c)

(c) Installation And Restoration. (1) The employer shall assure that all devices, components, combinations of devices, components, combinations of devices or systems constructed and installed to comply with this standard are approved. Steam whistles, air horns, strobe lights or similar lighting devices, or tactile devices meeting the requirement of this section are considered to meet this requirement for approval.

(2) The employer shall assure that all employee alarm systems are restored to normal operating condition as promptly as possible after each test or alarm. Spare alarm devices and components subject to wear or destruction shall be available in sufficient quantities and locations for prompt restoration of the system.

1910.165(d)

(d) Maintenance And Testing. (1) The employer shall assure that all employee alarm systems are maintained in operating condition except when undergoing repairs or maintenance.

Additionally, each employee or tenant-occupant should be made aware of the procedures for reporting emergencies, including the use of manual pull box alarms, public address systems, or telephones. Emergency phone numbers should be posted on or near telephones, as well as on employee and tenant-occupant bulletin boards, and in other conspicuous locations. The communications plan should be in writing; all employees and tenant-occupants should be made aware of its purpose and what action(s) should be taken in the event of an emergency or disaster.

Finally, it may be necessary to notify other key personnel such as the facility manager and/or backup person and medical provider during off-duty hours. An updated written priority-call list should be maintained at all times.

## Accounting for Personnel

Accounting for personnel is not only a critical issue it can also be a difficult issue to address. Accounting for people after an emergency site evacuation occurs can be compounded if the evacuation occurs during shift changes, or when other non-site personnel are occupying the building (e.g., contractors, subcontractors, vendors, clients, etc.). For these reasons, a designated member of the facility's emergency response team should be charged with this responsibility. Also, this person should have the additional responsibility and authority to inform the police, fire and appropriate medical personnel of any known or suspected missing people.

## The Emergency Response Team

The facility's emergency response team is the first line of defense in emergency and/or disaster situations. Because of the critical role that team members play, each person on the team must be both physically and emotionally capable of performing their assigned duties, which may include:

- Using various types of fire extinguishers.
- Performing first aid procedures, including cardiopulmonary resuscitation, or CPR.
- Conducting shutdown procedures.

- Implementing evacuation procedures.
- Using a self-contained breathing apparatus, or SCBA.
- Conducting search and emergency rescue procedures.

Since the type and extent of the emergency depends on site operations, the response will vary according to the type of process, materials involved, the actual number of employees and/or tenant-occupants and the availability of outside resources. Emergency response team members should be trained in all the possible emergencies and/or disasters that could occur at the site and in the appropriate response actions that must be performed. For example, team members should be informed about any special hazards that they many encounter during an emergency situation—such as the on-site location and storage of flammable materials, toxic chemicals, water-reactive substances, etc.

Equally important to note is that emergency response team members must also be advised of when not to intervene in an emergency or disaster situation. Certain crisis situations warrant professional firefighters and other professional emergency response personnel. The facility's emergency response team members must be able to determine when a situation is either too dangerous or too threatening to either themselves or to others. They must be intellectually astute enough to make those judgment calls that are often times required in emergency and/or disaster situations. In short, all emergency team members must temper enthusiasm and concern with common sense.

### Training

Training is the single most important factor in any emergency preparedness and disaster planning effort. Before any plan can be implemented, a sufficient number of people must be trained to assist in the safe and orderly evacuation of the site's personnel and occupants. Training for each type of disaster response is necessary so that emergency response team members know what actions are required.

Additionally, all site employees and tenant-occupants should be trained in:

- Evacuation plans.
- Alarm systems.
- "Safe-site" reporting procedures.
- Shutdown procedures.
- Types of potential emergencies and/or disasters.

Emergency and/or disaster training should be provided on an annual basis, minimally, and when:

- The plan is initially developed.
- New employees or tenant-occupants are assigned to the facility.
- New equipment, materials, processes, etc., are introduced into the facility.
- Procedures have been updated and/or revised.
- Test drills, rehearsals, etc., demonstrate the need for additional training.

When drills are conducted, it is advisable that they should include groups supplying outside services such as police and fire departments and medical standby facilities.

In addition to training, the company's written emergency control procedures should be made available to all facility employees and tenant-occupants. These control procedures, as well as the actual emergency preparedness/disaster and recovery plan, should be reviewed and updated periodically to maintain its intended objective of program efficiency.

**Personal Protection**

Effective personal protection is essential for people who may be exposed to potentially hazardous substances. This is particularly true for those members of a facility's emergency response team and/or for those people who remain to perform or shut down critical facility/plant operations before they evacuate.

In an emergency and/or disaster situation, people may be exposed to various hazards, including:

- Chemical splashes or contact with toxic materials.

- Falling objects and flying particles/debris, etc.
- Unknown atmospheres that may contain toxic gases, vapors, mists, etc., or inadequate oxygen to sustain life.
- Fires and electrical hazards.

To adequately protect people in these situations, various types of safety equipment may be required. These are:

- Safety glasses, goggles, or face shields for eye and face protection.
- Hard hats and safety shoes for head and foot protection.
- Respirators for breathing protection.
- Whole body coverings, as well as gloves, hoods and boots for protection from chemicals.
- Body protection for abnormal environmental conditions such as extreme temperatures.

Any safety equipment that is selected, however, should meet the standards set by ANSI and/or various regulatory agencies, including the National Institute for Occupational Safety and Health (NIOSH) and the Mine Safety and Health Administration (MSHA). And, any safety equipment that is, or will be used, should be selected *only after consultation* with safety and health professionals. For example, professional advice will most likely be needed in selecting appropriate respiratory protection. Respiratory protection is necessary for toxic atmospheres of dusts, mists, gases and vapors, as well as for oxygen-deficient atmospheres. There are four basic types of respirators:

1. Air-purifying devices, such as filters, gas masks, and chemical cartridges, which remove contaminants from the air but which cannot be used in oxygen-deficient atmospheres.

2. Air-supplied devices, such as hose masks, air line respirators, which should not be used in any atmospheres that are immediately dangerous to health or life.

3. Self-contained breathing apparatus, which are required for unknown atmospheres, oxygen-deficient atmospheres, or at-

mospheres, that are immediately dangerous to life or health (positive-pressure type only).

4.  Escape masks.

Before any member of the emergency response team uses any of the respiratory devices, however, it is imperative that the conditions shown below are satisfied:

1.  A medical evaluation must be made to determine if the people are physically able to use the device(s).

2.  Written procedures detailing the safe use and care of the devices must be developed prior to use.

3.  People who will be using the devices must be trained in the proper use and maintenance of the respirators.

4.  A "fit-test" must be made to determine the proper match between the face piece of the respirator and the face of the wearer. This testing must be repeated periodically. Training must also provide the wearer with opportunities to handle the respirator, have it fitted properly, test its face-piece-to-face seal, and wear it in both normal air for a familiarity period and in a test atmosphere.

5.  A regular maintenance program must be installed that includes cleaning, inspecting and testing of all respiratory devices. To ensure that respirators which are used for emergency response are in satisfactory working condition, the devices must be inspected after each use and at least monthly. A written inspection record must also be maintained.

6.  Distribution areas for equipment used in emergencies must be readily accessible.

SCBAs offer the best protection to people involved in controlling emergency and/or disaster situations. This apparatus should have a minimum service life rating of 30 minutes. Conditions that require SCBA include:

- Leaking cylinders or containers, smoke from chemical fires, or chemical spills that indicate high potential for exposure to toxic substances.

- Atmospheres with unknown contaminants or unknown contaminant concentrations, confined spaces that may contain toxic substances, or oxygen-deficient atmospheres.

Some emergency and/or disaster situations involve entering confined spaces to rescue people who are overcome by toxic compounds, or who lack oxygen. These confined spaces include tanks, vessels, pits, sewers, pipelines and vaults. Entry into a confined space can expose a person to a variety of hazards, including toxic gases, explosive atmospheres, oxygen deficiency, as well as electrical hazards and hazards that are created by mixers and impellers that have not been deactivated and locked out.

No person should ever enter a confined space under normal circumstances unless the atmosphere within the confined space has been tested for adequate oxygen, combustibility and toxicity. Unless proved otherwise, conditions in a confined space must be considered immediately dangerous to life and health. When a confined space must be entered because of an emergency, the following precautions must be taken:

1. All lines containing inert, toxic, flammable or corrosive materials must be disconnected or "valved-off" before entry.

2. All impellers, agitators, or other moving equipment inside the vessel must be locked out.

3. Appropriate personal protective equipment must be worn by people before they enter the confined space. Mandatory use of safety belts and harnesses must also be stressed.

4. Rescue procedures for each entry must be specifically developed.

When there is an atmosphere that is immediately dangerous to life or health, or a situation that has the potential for causing injury or illness to an unprotected person, a trained standby person should be present. The standby person should be assigned a fully-charged, positive-pressure, self-contained breathing apparatus with a full face piece. The standby person must also maintain unobstructed life lines and communications to all people within the confined space and be prepared to summon rescue personnel, if necessary. The standby person should not enter the confined space until adequate assistance is present. While awaiting rescue personnel, the standby person may make a rescue attempt utilizing life lines from outside the confined space.

**Medical Assistance**

In an emergency or disaster situation, time is a critical factor in minimizing injuries. There are a number of requirements that have been outlined by OSHA and which must be met to ensure availability of medical assistance. These requirements are:

1.  In the absence of an infirmary, clinic, or hospital in close proximity to the workplace that can be used for the treatment of all injured employees, the employer must ensure that an adequate number of employees are trained to render first aid.

2.  Where the eye or body of any employee may be exposed to injurious corrosive materials, eye washes or suitable equipment for quick drenching or flushing must be provided in the work area for immediate emergency use. Employees must be trained to use the equipment.

3.  The employer must ensure the ready availability of medical personnel for advice and consultation on matters of employee health. This does not mean that health care must be provided, but rather that, if medical problems develop in the workplace, medical help will be available to resolve them.

OSHA suggests that the following actions should be considered to fulfill the requirements which apply to all businesses regardless of size:

1.  Survey the medical facilities near the place of business and make arrangements to handle routine and emergency cases. A written emergency medical procedure should then be prepared for handling accidents with minimum confusion.

2.  If the business is located far from medical facilities, at least one and preferably more employees on each shift must be adequately trained to render first aid. The American Red Cross, some insurance carriers, local safety councils, fire departments, and others may be contacted for this training.

3.  First-aid supplies should be provided for emergency use. This equipment should be ordered through consultation with a physician.

4.  Emergency phone numbers should be posted in conspicuous places near or on telephones.

5.  Sufficient ambulance service should be available to handle any emergency. This requires advance contact with ambulance services to ensure they become familiar with plant locations, access routes and hospital locations.

**Security**

During a disaster, it is oftentimes necessary to secure a facility to prevent unauthorized access and to protect vital records and equipment. An off-limits area must be established by cordoning off the area with ropes and signs. In some situations it may become necessary to notify local law enforcement personnel to secure the area and to prevent the entry of unauthorized people to the facility.

Additionally, certain data may need to be protected, including employee and emergency number files, and legal and accounting fields. Companies and their individual facilities normally maintain duplicate copies of such data and information in separate protected locations.

## Summary

As incorporated into a company's disaster and recovery plan, these elements should address all potential emergency situations and/or potential disasters, whether the disaster results from a natural occurrence or through human error or design. These elements are also critical to the effectiveness of response time during an emergency, whether the emergency involves an accidental release of toxic gases, chemical spills, fires, explosions and any personal injury that may be sustained as a result of the occurrence.

## Sources

United States Department of Labor, Occupational Safety And Health Administration, *OSHA Handbook for Small Businesses,* Safety Management Series, OSHA 2209, 1992 (Rev.).

United States Department of Labor, Occupational Safety And Health Administration, *Principal Emergency And Preparedness Requirements in OSHA Standards And Guidance for Safety And Health Programs,* OSHA 3122, 1993 (Rev.)

Gustin, Joseph F., *Disaster and Recovery Planning: A Guide For Facility Managers, 2nd ed.* Atlanta: The Fairmont Press Inc., 2002.

# Chapter 6
# Emergency Response Model Plan

## USING THE MODEL PLAN

An emergency action plan must describe the procedures that the company/facility will develop and which must be communicated to and followed by all personnel in the event that an occurrence should necessitate the evacuation of personnel from the facility. The plan must also describe the various actions that management and employees will undertake to ensure safety from fire and other emergency occurrences.

This section provides the procedures and a sample plan that can be used by facility managers and other compliance professionals to customize, develop and implement a site-specific emergency response plan that can be used to satisfy the requirements of 29 CFR 1910.38(a). This sample Emergency Response Plan also addresses the provisions for employee alarm systems, as mandated by 29 CFR 1910.165. The sample plan addresses the basic elements of emergency planning as determined by OSHA:

- Types of evacuation to be used in emergency situations;

- Emergency escape procedures and emergency escape route assignments;

- Procedures to be followed by employees who remain to operate critical plan operations before they evacuate;

- Rescue and medical duties for those employees who remain to operate critical plant operations before they evacuate;

- The preferred means of reporting fires and other emergencies; and

- Names and regular job titles of persons and/or departments to be contacted for further information or explanation of duties under the plan.

Provision for training employees designated to assist in the emergency evacuation of employees is also included in this emergency action plan and is described in detail.

**Contents**
  Statement of Purpose
  Emergency Management Group
  Emergency Notification and Reporting
  Warning/Alarm Systems

## STATEMENT OF PURPOSE

This emergency action plan has been developed, written and installed in accordance with the applicable provisions of the Occupational Safety and Health Administration's Emergency Action Plan Standard 29 CFR 1910.38(a). The provision for employee alarm systems, as mandated by 29 CFR 1910.165, is also contained in this plan.

All site sections, units, departments, and divisions of (INSERT COMPANY/FACILITY NAME) are included in this Emergency Action Plan which will be available in the Facility Manager's office for review by all employees and/or their representatives.

The requirements of the referenced standards will be met by communicating emergency preparedness information to all site personnel and contractors through:

- Establishing an emergency management group to direct, coordinate and monitor emergency programming and procedures;

- Conducting comprehensive employee information and training programs;

- Installing and maintaining emergency alarm and warning systems; and

- Conducting periodic test drills and exercises, including evacuation drills.


## A.  EMERGENCY MANAGEMENT GROUP

1.  An emergency management group (EMG) has been established to direct, control and monitor all emergency occurrences and situations.

2.  The EMG is headed by the Facility Manager (IDENTIFY BY NAME) and is comprised of site personnel who have the authority to:

    i.   Determine both the long- and short-term effects of the emergency occurrence;

    ii.  Order a partial or total evacuation of the facility based upon the nature of the occurrence and its potential impact upon the immediate safety of site personnel;

    iii. Order the shutdown of the facility; and

    iv.  Coordinate activities with local jurisdictional authorities and other external organizations, including the media.

3.  A directory of EMG personnel is included in this plan and is identified as Exhibit I-a: EMG Directory.


## B.  EMERGENCY NOTIFICATION AND REPORTING

1.  Emergency notification and reporting procedures have been developed for all site personnel (A copy of same is included in this plan and identified as *Exhibit I-b; Emergency Notification and Reporting Procedures*).

2.  All personnel will be informed of emergency notification and reporting procedures at the time of hire/tenancy and periodically, thereafter.

3.  Site personnel designated to perform notification tasks will be briefed in appropriate methods and procedures for doing same. A copy of the notification briefing program, and names and job titles of site personnel designated to perform such tasks is included and identified as *Exhibit I-c: Notification Briefing*.

4.  A current directory of internal and external emergency response personnel and telephone and pager numbers is maintained by the EMG. Copies of same are distributed to all section, unit, department and division managers and supervisors. A copy of the directory is included in this plan and identified as *Exhibit I-d: Emergency Response Personnel Directory*.

5.  Emergency telephone numbers are posted near each telephone, on all employee/tenant bulletin boards, and in prominent locations throughout the facility.

## C.  WARNING/ALARM SYSTEMS

1.  Alarm systems have been installed to alert site personnel of an emergency situation. (NOTE TO USER: If your facility has either an audio or visual alarm, or a combination system, describe same. Also, if your facility does not have a combination alarm system, indicate what provision is made for warning personnel who are hearing or visually impaired).

2.  All site personnel are trained in procedures and measures to ensure appropriate response when alarm systems are activated. Training is provided at the time of hire/tenancy and periodically, thereafter.

3. An auxiliary power supply furnishes the necessary power to activate all warning/alarm systems, in the event that an occurrence renders the primary power source untenable.

4. Procedures have been developed for warning non-site personnel (e.g., contractors, customers, clients, etc.) of an emergency occurrence. A copy of these procedures is included in this plan and identified as Exhibit I-e: Emergency Warning Procedures for Non-Site Personnel.

5. All alarm systems are routinely inspected and tested on a monthly basis. Copies of inspection and testing results are included and identified as Exhibit I-f: Alarm Systems Inspections.

## D. FIRE SAFETY

1. Meetings with local fire authorities have been conducted to determine fire response capabilities and to familiarize the local fire department with facility operations, process and physical layout.

2. Site inspections will be conducted on a regular basis to determine the presence of fire hazards and/or situations that carry the potential for fire.

   Such inspections focus upon:

   - Identifying process and/or materials that could cause a fire, or contaminate the environment in the event of a fire; and

   - Complying with all applicable jurisdictional fire codes and regulations

3. Procedures for the safe handling and storage of all flammable liquids and gases have been developed and installed.

A copy of said procedures has been distributed to all employees whose job duties entail same.

4. Procedures to prevent the accumulation of combustible materials have been developed and installed. Copies of said procedures have been distributed to all supervisory and other personnel responsible for same.

5. All site personnel have been trained in appropriate response measures to fire alarm activation. Copies of training and employee briefings in response measures, employee names, job titles and schedule of training meetings is attached and identified as *Exhibit I-g: Fire Alarm Procedures—Employee Response Measure.*

6. Fire alarm systems are inspected on a monthly basis and maintained on an on-going basis.

7. As determined by code, smoke detectors have been placed in all appropriate locations throughout the facility. All smoke detectors are inspected on a monthly basis and batteries are replaced semi-annually, or as needed.

8. As determined by code, fire extinguishers have been placed in all appropriate and visible locations throughout the facility. All fire extinguishers are routinely inspected and maintained.

9. Site personnel have been trained in the appropriate handling and use of fire extinguishers. Training is conducted at the time of new employee/tenant orientation and at subsequent intervals as refresher training. Rosters of attendees and training meeting schedules are attached and identified as *Exhibit I-h: Fire Extinguisher Training.*
(NOTE TO USER: If your facility has installed a sprinkler and/or other fire suppression system, note same. Also, describe the frequency of inspection and maintenance of the suppression system.

10. Evacuation drills and exercises are conducted on a facility wide basis semi-annually, or as determined by local code.

11. Evacuation route maps are displayed in prominent locations throughout the facility.

12. Selected site personnel have been designated as fire wardens for their respective areas. All such personnel have been trained in procedures necessary for:
    - The safe evacuation of personnel within their respective areas; and
    - Monitoring shutdown procedures.

    A directory of internal fire wardens is included in this plan and identified as *Exhibit I-i: Directory of Internal Fire Wardens*.

13. Electrical power, gas, water utility shutoffs have been identified and clearly marked, to ensure responsive and timely shutdowns by fire wardens and/or external response units.

14. Safety training in basic fire prevention techniques is provided to all site personnel. Training topics include:

    - How to prevent workplace fires;
    - Reporting procedures;
    - How to contain a fire; and
    - How to evacuate.

    At the completion of the training session, class or meeting, all site personnel are required to verify in writing (by original signature) that they have attended the fire prevention training, and that they have received all information and training materials.

15. Selected site personnel from each area have been assigned rescue and medical duties. All such designated site personnel have received training in basic rescue and medical procedures, including cardiopulmonary resuscitation (CPR)

and assisting persons with disabilities, etc. All such training is conducted by approved providers. A list of designated site personnel and certified copies of training records are included in this plan and identified as *Exhibit I-j: Rescue and Medical Providers.*

16. Site personnel who perform rescue and medical duties have been trained in the proper use of personal protective equipment, including the proper use of the self-contained breathing apparatus.


**E.   EVACUATIONS**

1.   (IDENTIFY COMPANY/FACILITY NAME) has designated primary and secondary evacuation routes and emergency exits in the event that an occurrence requires the evacuation of site personnel.

2.   Designated evacuation routes have been determined in consultation with the local fire department. Evacuation plans have been coordinated with the local emergency management office. A copy of the company's detailed evacuation plan is included and identified as *Exhibit I-k: Evacuation Plan.*

3.   All designated evacuation routes and emergency exits are clearly marked and illuminated.

4.   Maps of evacuation routes have been distributed to all site personnel. Copies of same are included in *Exhibit I-k: Evacuation Plan.*

5.   The EMG has the authority to order evacuation of site personnel:

   - Depending upon the nature of the occurrence and the potential threat to the safety of site personnel, the EMG will order either a partial or total evacuation.

- Evacuation wardens have been designated to assist site personnel in the event of an evacuation and to ensure a safe and orderly evacuation. Evacuation wardens are responsible for accounting for site personnel within their respective areas.

- A copy of personnel accounting procedures is attached and identified as *Exhibit I-k: Evacuation Plan.*

6. Assembly Locations
   a. Locations have been designated where site personnel are required to assemble following an evacuation.

   b. Specific post-evacuation assembly areas are identified in *Exhibit I-k: Evacuation Plan.*

   c. Personnel accounting procedures will resume at the assembly location to verify safe evacuation of all site personnel.
      i. Names of persons not accounted for, and their last known locations, will be given to the EMG.
      ii. The EMG will notify rescue personnel of same.

7. **MAINTAINING VITAL OPERATIONS**

   a. Selected site personnel have been designated to maintain vital facility operations in the event of an emergency/disaster occurrence.

   b. Designated personnel have been trained in
      i. Methods and procedures for maintaining such operations;
      ii. Proper use of personal protective equipment when and where necessary; and

    iii.   Methods and procedures for aborting such operations to ensure their own safe evacuation.

    c.   A list of all such designated personnel and the training program outline is included in this plan and identified as *Exhibit I-l: Vital Operations.*

**Sources:**

Gustin, Joseph F. *The Safety Emergency Health Planning* Disk. New York: UpWord Publishing, Inc., 1996.

# Chapter 7
# *Communications Systems*

The communication system is the hub of the office. The components that consist of the system—the telephone, computer and video system—entail a complex distribution network that has a significant impact on occupied space. Satellite desks, microwaves, fiber optics and telephone lines make up this distribution network. Each of the basic types of cabling—workstation, riser and in the case of multiple buildings, plat cabling—occupy much needed space.

## MINIMUM SPACE REQUIREMENTS

Minimum space recommendations from the Telecommunications Industry Association (TIA) allocates a minimum of one closet per floor and a minimum of at least 70 square feet for each 10,000 square feet of floor space. Multi-vendor/multi-technology spaces may even need more space.

Having to accommodate these network distributions include wire distribution through dropped ceiling space which serves either the floor above or below in multi-story facilities; under raised flooring or imbedded in the floor structure.

According to the General Services Agency, (the GSA), this floor distribution, especially when it is integral to the structural system, can provide advantages in flexibility and, if combined with mechanical systems, can reduce the amount of horizontal circulation space required for utilities.

## TYPES OF DISTRIBUTION SYSTEMS

The GSA lists the major types of distribution systems that impact upon building infrastructure. These distribution systems include the wiring for power, data and building-control systems.

These major types of distribution include:

- Hardwired systems

- Zoned systems

- Wireless technologies

- Plug-and-Play networks

"Hardwired" systems are those where individual workstations are connected to a telecommunications closet by an uninterrupted length of cable. This is the least flexible and most cumbersome system, using a considerable amount of space for "home-run" cabling.

"Zoned" systems are where workstations are clustered in groups of up to about 25, with server boxes arranged on a grid beneath an access floor. Since only the boxes are connected to the telecommunications closet the amount of cabling used is reduced.

Some of the disadvantages of the Zoned system include:

*Immobility*

A user's computer or telephone cannot be mobile in a zoned underfloor system. "Plugging in" a computer or telephone from one set of outlets to another cannot be done because each move necessitates the reconfiguring of the outlets.

*Inaccessibility*

Distribution boxes need to be accessible on a routine basis. Therefore this system is unsuited for use above ceilings or mounted to walls.

### Difficulty in Tracking Problems

The greater the number of connection points in a zoned system make changes and problem tracing more difficult to track.

### Lack of Portability

LANs cannot accommodate much movement of computer equipment because of limitations in the way networks are structured. Office administrators discourage portability because it increases equipment maintenance and reconfiguration costs.

## "Wireless" Technologies

"Wireless" technologies provide connections with, as the name suggests, no wires to the equipment. The advantages of this type of technology include:

### Portability

Computers and telephones can be moved from one location to another without having to be reconfigured with each move as in a "zoned" system. The restrictions of wireless technologies include:

### Slow Data Transmission

Since data transmission is slower in an office as compared with wired systems, they are especially useful outside the office setting—convention centers or utility areas are ideally suited for such configurations since this setting demands a high need for portability and mobility coupled with a relatively low data demand.

Where the mobility of wireless outweighs its disadvantages, provision must be made for an above-ceiling "grid" of antennas and wire pathways to service those antennas.

### Limited Ranges

Wireless systems have a limited range of radio frequency and infrared spectrum. Once these frequencies have been exhausted, these systems slow down or completely stop. This system also is encumbered by airborne signal encryption requirements.

*Fixed Space Requirements*

Wireless technologies can complicate flexibility where workstations require hard wiring for power and task lighting. In many cases space changes become complicated because system performance is affected by furniture and wall locations.

*Building Exterior Maintenance/Appearance*

Since antennas are required for wireless technologies, the placement of antennas impact the building's exterior.

## "Plug-and-Play"

These networks allow computer and telephone network access from any desk, but have the same limitations as the wireless communications. The technical and data handling capabilities are limited in an office setting.

## Wire Handling Options

The wire handling options for a building's communication system include:

- "Poke-through" wire handling.
- Ceiling and wall-based distribution systems.
- Access flooring.

*"Poke-through" Wire Handling*

This type of distribution is an extension of the traditional electrical distribution system. Users are served through a floor penetration, by a system located in the ceiling of the floor below. Since the building's integrity can be affected by holes drilled through slabs and patching previously made holes, this system is best suited to spaces with low device density, such as a lobby.

*Ceiling and Wall-based Distributions Systems*

These are the most commonly used systems. They are economical. They provide enough space for a large number of cables and large bending radii of copper cable and optical fiber. They can be moved with less frequent damage because of their use of support trays with access boxes at frequent intervals. Most often, cable trays are located in halls with conduit stub-ups from workstation outlet boxes. Disadvantages of this system include tracing problem cabling and maintaining a clear space above the ceiling

*Access Flooring*

Access flooring is used extensively in main computer rooms for electrical, data and conditioned air distribution. Thinner access-floor tiles and thinner system wires commonly available allow the use of floors as little as 1-1/2 inches high for wire distribution. When using access flooring, transitions to other building elements such as ramps, elevator offsets, and slab depressions must be accommodated.

**Rooftop Systems**

As stated earlier, rooftop systems offer its own challenges. The Occupational Safety and Health Administration (OSHA) and the Federal Communications Commission (FCC) rules must be followed to minimize safety and liability concerns.

Building owners and managers have a responsibility to maintain an environment that ensures the safety of workers, tenants and the general public.

When considering using rooftop systems for a building's communication network, building owners and facility managers should be aware that the antennas used or installed should be checked for emissions to ensure that they do not exceed the Maximum Permissible Exposure—MPE—as determined by the Federal Communications Commission (FCC). Experienced Radio Frequency (RF) engineers should be able to assist. If MPE limits are exceeded, then OSHA enters the picture. OSHA mandates procedures to mitigate any potential hazards posed by RF emissions.

**Sources**

U.S. General Services Administration, Office of Real Property, Office of Governmentwide Policy *New Adventures in Office Space: The Integrated Workplace*, A Planning Guide. February 2002.

U.S. General Services Administration, Office of Real Property, Office of Governmentwide Policy. *The Integrated Workplace: A Comprehensive Approach to Developing Workspace*. Second Printing April 2000.

# Chapter 8
# The Americans with Disabilities Act

Perhaps no other legislative act in recent history has impacted facility management as much as the Americans with Disabilities Act (ADA). Its scope is far reaching and pervades every aspect of facility management. From the employment provisions (Title 1) of the ADA, to the Public Accommodations and Commercial Facilities provisions (Title III), the ADA defines, in large part, the role of facility manager as compliance officer.

The areas presented in this discussion of the ADA are those that have particular relevance for facility managers. The significance of the Title I employment provisions cannot be underestimated. Title I determines, in large part, the approach facility managers must take to meet their responsibilities to ensuring worker safety and in managing the safety function.

Title II, which governs state and local government operations, as well as Title III, which covers private entities that have been determined to be places of public accommodation or commercial facilities, carry additional responsibilities for facility managers in both the public and private sectors.

These responsibilities include the facility managers' obligations to maintain environments that ensure the safety of people with disabilities, but also minimize risk of injury to those employees who provide services to people with disabilities.

## BACKGROUND

The ADA is a comprehensive civil rights law for people with disabilities. The Equal Employment Opportunity Commission (EEOC) enforces the ADA's Title I employment practices by private entities. The Department of Justice enforces the ADA's requirements in three areas:

Title I:  Employment practices by units of state and local governments

Title II:  Programs, services and activities of state and local governments

Title III:  Public accommodations and commercial facilities of private entities.

### Title I: Employment

The ADA was signed into law on July 26, 1990. The employment provisions of the Act, referred to as Title I, became effective on July 26, 1992 for employers with 25 or more employees. July 26, 1994 marked the effective compliance date for employers with 15 to 24 employees.

The ADA prohibits employers from discriminating against qualified individuals with a disability who, with or without reasonable accommodations, can perform the essential functions of a job. This prohibition extends to:

- The job application process.

- Any term of employment, including hiring, advancement, discharge, compensation, job training, and any and all other terms, conditions and privileges of employment.

The intent of the ADA is to ensure access to employment based on merit. The ADA does not create preferences in favor of individuals with disabilities, nor does it establish quotas or affirmative action requirements.

**Title II: Programs, Services and
Activities of State and Local Government**

Title II provisions of the ADA cover "public entities" which include the "state or local government and any of its departments, agencies, or other instrumentalities."

The specific focus of Title II is on any and/or all activities, services and programs of public entities that include the activities of:

- State legislatures and courts.
- Town meetings.
- Police and fire departments.
- Motor vehicle licensing.
- Employment.

Municipally operated public transportation systems, as well as other state and local government-operated transportation systems are covered by Department of Transportation regulations.

These regulations establish specific requirements for transportation vehicles and facilities, including a requirement that all new buses must be equipped to provide services to people who use wheelchairs.

**Title III: Public Accommodations and Commercial Facilities**

Referred to as Title III of the ADA, the provisions concerning public accommodations and commercial facilities became effective on January 26, 1992. The significance of Title III is twofold:

1. Title III mandates that public accommodations and commercial facilities be readily accessible to people with disabilities.

2. It directly impacts upon those people who are responsible for the operation and use of facilities, including landlords, tenants, owners, operators and facility managers.

This mandate directly impacts upon facility managers because Title III regulations cover:

- Private entities that own, operate, lease from or lease to places of public accommodation.
- Commercial facilities.
- Private entities that offer certain examinations and courses related to educational and occupational certification.

Like the Title I employment provisions and the Title II provisions which regulate programs, services and activities of state and local government, Title III is a civil rights law that prohibits discrimination against individuals with disabilities. Its purpose is to promote the accommodation of people with disabilities in the delivery and receipt of goods and services.

## TITLE I AND THE FACILITY MANAGER

Unlike the standards that regulate state and local government programs, services and activities, or public accommodations and commercial facilities (Title II and Title III, respectively), the employment provisions of ADA's Title I are less "clear cut" and definitive.

For the facility manager who is responsible for integrating the employee with the work environment, this issue is compounded by three different, but interrelated, factors.

First, the preponderance of disability-related discrimination complaints filed with the Equal Employment Opportunity Commission (EEOC), which enforces Title I of the ADA, are work-injury related.

Second, confusion surrounds an injured worker's "status" as an individual with a disability. While the ADA does not specifically identify an injured worker as a person with a disability, it does not exclude an injured worker from its definition of a person with a disability. The key to determining the injured worker's "status" rests with the worker, meeting the ADA's definitional test. If the injured worker does meet ADA criteria, that individual assumes the same rights as other qualified individuals under the ADA.

Third, the impact of workers' compensation upon the injured worker adds to the complexity, especially in terms of benefits entitlement, fitness to return to work with or without reasonable accommodation, and the employer's ability/inability to reasonably accommodate the injured worker with a disability.

## The ADA's Definition of Disability

The ADA defines a person with a disability as an individual who:

- Has a physical or mental impairment that substantially limits one or more major life activities.
- Has a "record" of such an impairment.
- Is regarded as having such an impairment.
- Is associated with an individual with a disability.

## Physical and Mental Impairments

A physical impairment is defined as any physiological disorder or condition, cosmetic disfigurement or anatomical loss affecting one or more of the following body systems: neurological, respiratory (including speech organs), cardiovascular, reproductive, digestive, genital-urinary, hemic and lymphatic, skin and endocrine.

A mental impairment is defined as any mental or physiological disorder, such as mental retardation, organic brain syndrome, emotional or mental illness, and specific learning disabilities.

The ADA does not provide a comprehensive list of physical and mental impairments. The number and variety of impairments preclude it from doing so. The impairment itself, however, must either be either physiological or mental. Simple physical characteristics such as height/weight within a normal range, left handedness, hair/eye color, etc., are not impairments.

Determination of impairment is made without regard to any kind of medication or device that an individual may be required to take or use. A person required to use a vasodilator to control asthmatic attacks, for example, is considered to have an impairment, even if the use of the vasodilator completely prevents attacks.

Persons who have contracted contagious diseases such as tuberculosis, or Human Immunodeficiency Virus (HIV), are considered to have an impairment under the ADA.

While the ADA recognizes drug addiction as an impairment, employees (and applicants) are not protected from any personnel actions based on current use of illegal drugs. Individuals who are no longer illegally using drugs and who have been successfully rehabilitated, or are in the process of completing a rehabilitation program, however, are protected under the Act.

Stress and depression may be considered impairments under EEOC guidelines. Determination is based upon documented physiological or mental assessments.

*Major Life Activities*

The impairment must substantially limit one or more major life activities for the individual to be considered as having a disability under the ADA.

Major life activities are defined as those basic activities that the average person in the general population can perform with little or no difficulty. EEOC Guidelines list the following as representative of major life activities:

| walking | speaking | breathing |
| seeing | hearing | sitting |
| standing | reaching | lifting |
| reading | manual tasks | working |
| caring for one's self | | |

*Substantially Limiting*

Under the ADA, an impairment is a disability if it "substantially limits" one or more major life activities. There are three factors that are used to determine if an impairment substantially limits a major life activity:

- The nature and severity of the impairment.
- The duration/expected duration of the impairment.
- The long-term or permanent impact/or expected long-term or permanent impact of the impairment.

It is important to consider these, because it is the *effect* of the impairment that determines if one or more of a person's major life activities has been substantially limited. The same holds true for the individual who has multiple impairments.

While no one impairment may substantially limit a major life activity, the cumulative effect of those impairments may limit a major life activity. When and if this occurs, the individual does have a disability.

*Has A "Record" of Such an Impairment*

Under the ADA, people who have a history of disability are protected, whether or not they are currently substantially limited in a major life activity. People who have either been wrongly classified or misdiagnosed as having a disability are also protected under the ADA.

*Is Regarded as Having an Impairment*

Perception of impairment is a significant factor. If a person is perceived to have an impairment that substantially limits a major life activity, then that person is protected under the ADA.

**Temporary Impairments**

The duration of an impairment is not the only factor that determines substantial limitations.

The extent of the impairment, as well as the impact, or effect of the impairment are factors that must be considered. A temporary, non-chronic impairment that does not last for any given length of time, and that does not have a long-term impact upon the individual, is not considered a disability for the most part.

**Qualified Individuals with Disabilities**

For an individual to be protected by the ADA, the individual must be qualified as such. In other words, a qualified individual with a disability is a person who meets the prerequisite skills, experience, educational background and all other requirements of the job, and who can perform the essential functions of the job, with or without reasonable accommodation.

If the person meets the necessary prerequisites of the job, then the person is considered "otherwise qualified."

In this case, the hiring manager must determine if that person can perform the essential functions of the job with or without reasonable accommodations.

In addition to their own evaluation, or judgment as to whether or not the person can perform the essential functions of the job, the hiring manager must consider:

- The job description (job descriptions must accurately reflect the functions of the job the employee will be performing).
- The actual amount of time the person will spend performing the functions of the job.
- Any consequences of not requiring the person to perform the functions of the job.
- The work experience of people who have performed the job in the past, as well as persons who are currently doing the job.
- The nature of the work operation.
- Collective bargaining agreements (if applicable).

If the individual cannot perform the essential function of the job, with or without reasonable accommodation then that person is not considered qualified by the ADA.

**Reasonable Accommodation**

The ADA requires that employers reasonably accommodate current employees and job applicants with disabilities. The general rule states than an employer cannot select a qualified person without a disability over an equally qualified person with a disability only because the employee/applicant with a disability will require accommodation.

Some examples of accommodation include making existing facilities accessible to people with disabilities, job restructuring, equipment and/or worksite modification, provision of assistance aids (e.g., Braille materials, etc.) and job reassignment if the person can no longer perform his original job duties.

Because reasonable accommodation must be made on an individual basis, the EEOC holds that reasonable accommodation must be jointly determined by employer and employee (or applicant). This is accomplished by consulting with the qualified employee or applicant with a disability to identify limitations, as well as appropriate agencies, rehabilitation organizations and medical care providers to determine potential accommodations. The accommodation appropriate for the person to perform the essential functions of the job is then chosen.

Note that while the accommodation must be reasonable, it does not need to be the best. If a less expensive assistance aid (e.g., manually-operated wheelchair) will permit the person to perform the essential job functions, for example, the employer does not need to provide a better or more expensive aid (e.g., motorized wheelchair). The employee (or applicant) may, however, choose a different accommodation if he is willing to provide it.


## TITLE II AND THE FACILITY MANAGER

The significance of Title II rests with the provision of programs, services and activities by state and local government entities.

With the exception of public transportation services that are operated by state and local governments, any public entity—including departments, agencies, or other divisions of state and local government—are subject to Title II provisions.

(Note: The United States Department of Transportation's (U.S. DOT's) regulations establish specific requirements for transportation vehicles and facilities, including a requirement that all new busses be equipped to provide services to people who use wheelchairs.)

### Compliance with Title II

State and local governments are required to:

- Permit any and/or all persons with disabilities to participate in any and/or all programs, services or activities under the entity's jurisdiction.

- Provide programs and services in an integrated setting, unless separate or different measures are necessary to ensure equal opportunity.

- Eliminate unnecessary eligibility standards or rules that deny individuals with disabilities equal opportunity to enjoy the programs, services or activities under the jurisdiction of the public entity, unless doing so is "necessary" for the ongoing provision of the program, service or activity; as well as any requirement that ends to screen out individuals with disabilities is prohibited; and any safety program, however, such as eligibility for drivers' licenses, may be imposed if they are based on actual risks and not on mere speculation, stereotypes or generalizations about individuals with disabilities.

- Furnish auxiliary aids and services, when necessary, to ensure effective communication, unless an undue burden or fundamental alteration would result.

- Make reasonable modification to policies, practices and procedures that deny equal access to individuals with disabilities, unless a fundamental alteration to the program would result.

- Operate their programs so that, when viewed in their entirety, the programs are readily accessible to and usable by individuals with disabilities.

Public entities are also prohibited from placing special charges on individuals with disabilities to cover the costs of measures necessary to ensure nondiscriminatory treatment, such as making modifications required to provide program accessibility or providing qualified interpreters.

Generally speaking, however, the areas that have the greatest impact upon facility managers are program access, including integrated programs, communications and new construction and alterations. Each of these areas, however, is addressed in terms of the eligibility requirements of the individual with a disability.

**Eligibility Criteria**

To repeat, an individual with a disability is a person who has a physical or mental impairment that substantially limits a "major life activity"; has a "record" of such an impairment; is regarded as having such an impairment; or is associated with an individual with a disability.

The "qualified" individual with a disability is a person who meets the essential eligibility requirements for the program, service or activity offered by the public entity.

"Essential eligibility requirements" depend on the type of program service or activity that is involved. In some activities, such as a state licensing program, the ability to meet specific skill and performance requirements may be "essential." The "essential eligibility requirements" would be minimal in places where the public entity provides information to anyone requesting it.

Also note that although the ADA recognizes drug addiction as an impairment, state and local government entities may withhold any services or benefits to an individual who currently engages in the use, possession or distribution of illegal drugs. In making a determination to withhold services and benefits, the public entity must carefully review all facts to ensure that its belief is reasonable. Individuals who have completed a supervised drug rehabilitation program, or who are currently participating in a supervised drug rehabilitation program and who are not otherwise engaged in illegal drug activity, are protected under Title II of the ADA.

**Program Access**

Under Title II provisions, state and local governments are required to:

- Ensure that persons with disabilities are not excluded from programs, services and activities because buildings are inaccessible.

- Make the programs accessible to individuals who are unable to use an existing facility that is inaccessible.

- Provide alternative delivery methods for the programs, services and activities currently offered in an inaccessible building (e.g., relocating a service to the first floor of a building; providing personal assistance to enable a person with a disability to obtain the service; providing benefits or services at the individual's home or at an alternative accessible site).

State and local government entities are not required to take any action that would result in a fundamental alteration in the nature of their programs, services or activities. Nor are they required to take any action that would result in undue financial and/or administrative burdens. However, public entities are required to take any available action that would ensure that persons with disabilities would receive benefits or services, provided that the action would not result in fundamental changes or create undue burden.

## Integrated Programs

Integrating persons with disabilities into the mainstream of society is fundamental to the ADA. For this reason, public entities are generally prohibited from offering programs and services that are separate or different from programs and services offered to people without disabilities. Public entities may offer separate programming or services in those instances where separate programming services are necessary to ensure that benefits or services are equally effective, however.

In those instances where a separate program is permitted, the person with a disability still maintains the right to participate in a regular program.

State and local governments may not require any person with a disability to accept an accommodation or benefit. The person with a disability retains the right to refuse accommodation or benefit.

## Communications

State and local governments are required to ensure effective communications with persons who have disabilities. To accomplish this, the public entity must provide appropriate auxiliary or assistance aids.

Examples of auxiliary aids, which must be provided at no cost to the person with a disability, include (but are not limited to):

| | |
|---|---|
| Qualified interpreters | Readers |
| Assistive listening devices | Taped text |
| Televised captioning and decoders | Braille materials |
| Large print materials | TDDs |
| Videotext displays | |

## New Construction and Alterations

State and local governments must ensure that newly constructed buildings and facilities are free of architectural and communication barriers that restrict access or use by persons with disabilities.

While the ADA does not require retrofitting of existing buildings to eliminate barriers, it does require the entity to ensure accessibility at the time any alteration is undertaken.

### *Technical Standards for Accessible Design*

Since its enactment, the ADA has provided public entities with the option of choosing between two technical standards for accessible design: The Uniform Federal Accessibility Standard (UFAS), established under the Architectural Barriers Act; and the Americans with Disability Act Accessibility Guidelines (ADAAG), adopted by the U.S. Department of Justice (DOJ) for places of public accommodation and commercial facilities covered by Title III of the ADA.

## Enforcement

State and local government entities are subject to the Title II provisions of the ADA, which are enforced by the U.S. Department of Justice. Individuals, as well as classes of individuals who believe they have been discriminated against by any state or local government entity may file complaints with any federal agency that provides financial assistance to the state or local program in question or with the U.S. Department of Justice, which will refer the complaint to the appropriate agency.

Remedies available are the same as provided under section 504 of the Rehabilitation Act of 1973. Reasonable attorneys' fees may also be awarded to the prevailing party.


## TITLE III AND THE FACILITY MANAGER

Title III of the ADA mandates that public accommodations and commercial facilities be readily accessible to persons with disabilities. This mandate directly impacts many facility managers because Title III regulations cover:

- Private entities that own, operate, lease from or lease to places of public accommodation.

- Commercial facilities.

- Private entities that offer certain examinations and courses related to educational and occupational certification.

Like Title I and Title II provisions of the ADA, Title III is a civil rights law that prohibits discrimination against people with disabilities. Its purpose is to promote the accommodation of persons with disabilities in the delivery and receipt of goods and services.

### Individuals with Disabilities

Again, an individual is considered as having a disability if he has a physical or mental impairment that substantially limits a "major life activity," has a "record" of such an impairment and is regarded as having an impairment, whether he has one or not.

Also included in the definition of impairment is drug addiction. In most cases, however, a public accommodation may make a decision to withhold services or benefits to an individual who currently engages in the use, possession or distribution of illegal drugs.

"Current use" is the illegal use of controlled substances that occurred recently enough to justify a reasonable belief that a person's drug use is current, or that continuing use is a real and ongoing problem. In basing a decision to withhold services or benefits from a person engaged in illegal drug activity, the private entity must carefully review all of the available facts to ensure that its belief is reasonable.

People who have successfully completed a drug rehabilitation program, or are currently participating in a supervised drug rehabilitation program and are not currently engaging in illegal drug use, are protected by Title III of the ADA.

**Defining Public Accommodations and Commercial Facilities**
*Public Accommodations*

A facility is considered a place of public accommodation when it is operated by a private entity, its operation affects commerce and it falls within one of the categories determined by the ADA to be a place of public accommodation:

- Places of lodging such as hotels and motels, except those owner-occupied places that rent fewer than six rooms.

- Restaurants, bars and other eating and drinking establishments.

- Movie theaters, stadiums, concert halls and other places of entertainment.

- Auditoriums, convention centers, lecture halls and other facilities where people gather.

- Retail and service facilities, such as banks, lawyers and accountant offices, hospitals, offices of health care providers, barber and beauty shops, pharmacies and insurance offices.

- Public transportation terminals, depots and stations with the exception of facilities related to air transportation.

- Libraries, galleries, museums and other places of public display or collection.

- Recreational areas such as amusement parks, zoos and parks.

- Private educational facilities, including graduate and under-graduate schools, secondary, elementary and private nursery schools.

- Social service facilities including senior citizen centers, child day care and homeless centers, food banks and adoption agencies.

- Exercise facilities including golf courses, health spas, bowling alleys and gymnasiums.

In short, a public accommodation is any place that involves the general public and the general public's right to buy, sell or engage in a variety of activities within the facility.

*Facilities Not Considered Public Accommodations.* There are several places of public gathering that are not considered to be places of public accommodation.

These include any residential apartment/condominium complexes which are subject to Fair Housing Act provisions, religious entities and private clubs. (Religious entities may be subject to the Title I provision of the ADA if their congregations employ the number of persons required for coverage. Private clubs are exempt only if they meet the specific criteria defined in Title II of the Civil Rights Act of 1964).

*Defining Commercial Facilities*

Commercial facilities, as defined by the ADA, are facilities that are intended for non-residential use by a private entity, and whose operations affect commerce.

Commercial facilities include warehouses, factories, office buildings, airports and wholesale facilities that sell exclusively to businesses.

*Facilities Not Considered Commercial Facilities.* Generally speaking, a commercial facility involves only a select group of individuals. For example, only a company's employees and business associates are authorized to engage in activities within the facility. Any facility that is covered by the Fair Housing Act of 1968, such as residential apartment complexes, are specifically exempted from the ADA's definition of commercial facilities.

### Combined Facilities

Companies that operate as a "mixed-use" or combined facility are subject to Title III provisions as well. A company that operates a warehouse that sells goods to a specific trade, for example, is not considered a public accommodation. However, those companies which sell their goods/products to the general public in any company-operated outlet store or franchise must comply with the public accommodations provision of the ADA.

## Public Accommodations: Employer Responsibilities

Public accommodations under the ADA carry specific obligations for employers. These obligations include:

- Removing architectural barriers and communications barriers that are structural in nature, to the extent that it is "readily achievable." Cost and the necessity for making the change(s) determine if removal of barriers is readily achievable.

- Determine if other alternatives are available when barrier removal is not readily achievable.

- Provide for auxiliary aids and/or services to persons with disabilities when and where necessary to prevent those persons from being excluded, denied service, segregated or otherwise treated differently than the general population.

- Make reasonable modifications to policies, practices and procedures that are necessary to providing equal goods and services to persons with disabilities.

- Review any eligibility requirements or admissions criteria to ensure that persons with disabilities are not excluded.

*Removal of Architectural Barriers*

A facility that meets the ADA's definition of public accommodation must remove architectural barriers when it is readily achievable. While there is no requirement to remove a barrier from an area that is accessible only to employees, the area itself may fall under the regulatory scope of Title I employment provisions.

*Priorities for Barrier Removal.* The U.S. Department of Justice, which enforces Title III provisions of the ADA, has determined priorities for removal of barriers. The purpose of these priorities is to facilitate long-term business planning and to maximize the degree of effective access that will result from any given level of expenditure.

The priorities are not mandatory. Public accommodations have the latitude to determine the most effective "mix" of barrier removal measures to undertake in their facilities.

The first priority is to enable persons with disabilities to physically enter the facility. This priority is generally preferable to other alternative arrangements in terms of both business efficiency and the dignity of persons with disabilities.

The second priority is for measures that provide access to those areas of a place of public accommodation where goods and services are made available to the general public. To the extent that it is readily achievable to do so, persons with disabilities should be given access to front desk assistance in a store, as well as access to the retail display areas of the store, for example.

The third priority should be providing access to restrooms, if restrooms are provided for use by customers and clients.

The fourth priority is to remove any barriers to using a place of public accommodations facility (for example, lowering public telephones).

*"Readily Achievable."* For an architectural barrier to be "readily achievable," it must be "easily accomplishable and able to be carried out without much difficulty or expense." Factors that must be considered in removing a barrier are the nature of the barrier and its cost, the financial resources of the site involved, the distance of the site from the parent company or other entity, the financial resources of these parties, and the type of business operation of these parties.

If the barrier removal is not readily achievable, the facility must make its goods and services available to persons with disabilities through alternative methods. Any measures taken to provide alternative methods, however, cannot pose a significant risk to the safety or health of persons with disabilities, or others. (See Appendix II "Checklist for Existing Facilities" for a listing of possible solutions to barrier problems.)

When a facility determines that installing a permanent ramp to accommodate persons who use wheelchairs is not readily achievable, for example, a portable ramp may be provided. To ensure safety, the portable ramp should have railings and a firm and stable non-slip surface. The portable ramp should also be properly secured.

*Provide Auxiliary Aids*

Facilities that meet the definition of public accommodations must provide auxiliary aids to ensure equal accessibility unless an undue burden or fundamental alteration would result.

Various types of auxiliary aids that must be provided include:

Qualified interpreters
Telecommunications devices for deaf persons (TTDs)
Closed captioned decoders
Qualified readers for sight-impaired persons
Braille materials
Audio tapes
TTDs for persons with speech impairments
Computer terminals
Speech synthesizers

*"Undue Burden."* Public accommodations are not required to provide auxiliary aids if doing so would result in an undue burden. The same criteria used to determine the feasibility of removing barriers apply to making an undue burden determination.

*"Fundamental Alteration."* Public accommodations are not required to provide auxiliary aids for people with disabilities, if a fundamental alteration in the nature for the goods and/or services would result.

*Modify Policies To Avoid Discrimination*

Facilities that are public accommodations must reasonably modify policies, practices and procedures to avoid discriminating against people with disabilities. Sight-impaired persons who use service animals (i.e., guide dogs), for example, must be permitted entry into a public accommodation, unless doing so would result in either a fundamental alteration, or if the safe operation of the facility would be jeopardized.

Public accommodations are prohibited from charging a fee to persons with disabilities to cover the cost of the accommodation. For example, a fee for home delivery of goods and/or services may not be charged to persons with disabilities, unless the general population is charged the same fee for home delivery.

*Ensure Non-Discriminatory Eligibility Criteria*

Public accommodations are prohibited from establishing eligibility criteria that tend to screen out persons with disabilities. A violation of Title III provisions would result, for example, if a bank required a valid driver's license as proof of identity for new depositors.

Public accommodations are also prohibited from making unnecessary inquiries into the existence of a disability. A private summer camp, for example, may require parents to complete a questionnaire and to submit medical documentation regarding their child's ability to participate in the various activities. The questionnaire, itself, is acceptable if the camp can demonstrate that each piece of information requested is needed to ensure the child's safe participation in activities. The camp is prohibited from using the medical information to screen out children from the camp, however.

Places of public accommodation may impose any legitimate safety requirements that are necessary for safe operation. A wilderness tour company may require participants to meet a certain level of swimming proficiency to participate in a rafting expedition, for example.

## Constructing And Renovating Commercial Facilities

The ADA's Accessibility Guidelines (ADAAG) determines design and construction standards, as well as accessibility standards for alterations and renovations to exiting facilities. These standards, which went into effect in January 1992 for facility renovation, and January 1993 for new construction, are precise (See Appendix I, "ADA Checklist for New Lodging Facilities" for specific guidelines. While these guidelines are specifically designated for New Lodging Facilities, many of the items are useful and applicable for alterations and renovations to many buildings.)

*Alterations And Renovations*

Alterations to existing public accommodations and commercial facilities begun after January 26, 1992 must be readily accessible to and usable by disabled persons, to the maximum extent feasible.

Any alteration to a facility's primary function area must assure that paths to restrooms, telephones and drinking fountains are accessible to and usable by persons with disabilities. The only exception is when the costs of altering primary function areas to provide accessibility exceeds 20 percent of the total alteration costs. If this occurs, however, the facility must still expend 20 percent of the total alterations cost, giving priority to those elements that provide the greatest degree of access. Changes should be made in the following order: accessible entrance, accessible route to the altered area, at least one accessible restroom for each sex (or single unisex restroom), phones, drinking fountains, and then other elements such as parking, storage and alarms. (See Figures 8-1 through 8-13).

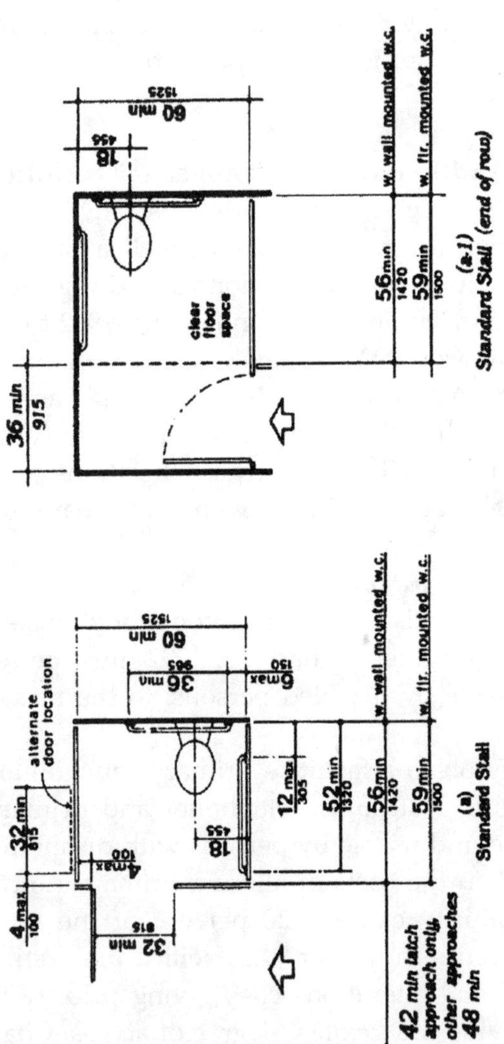


Figure 8-1 (facing pages). Toilet Stalls
(Courtesy: United States Department of Justice)

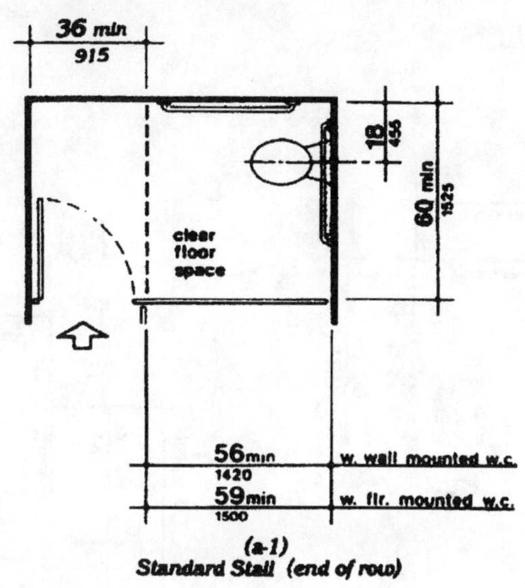


(a-1)
Standard Stall (end of row)

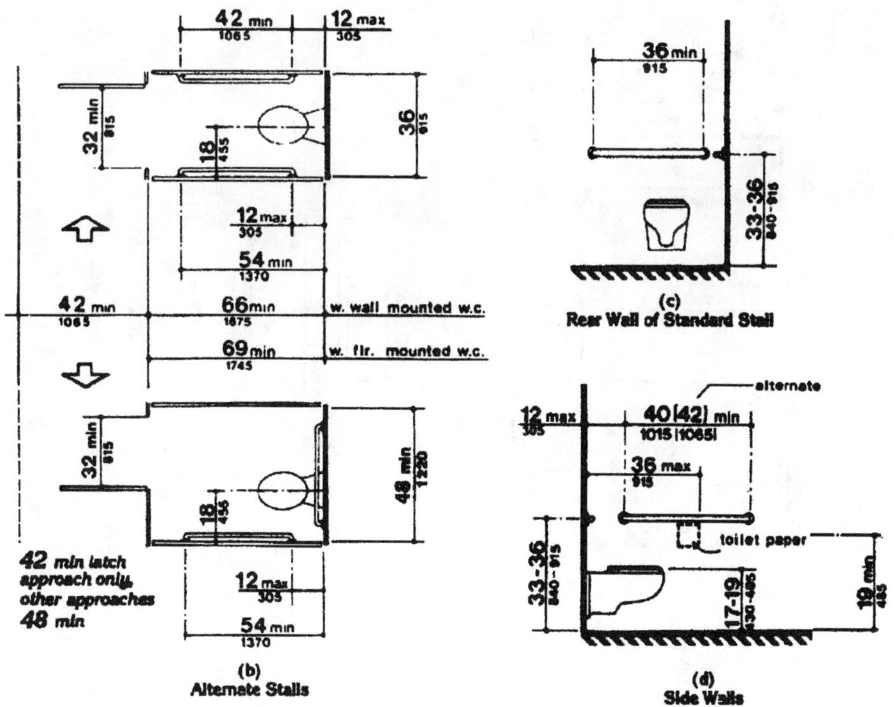

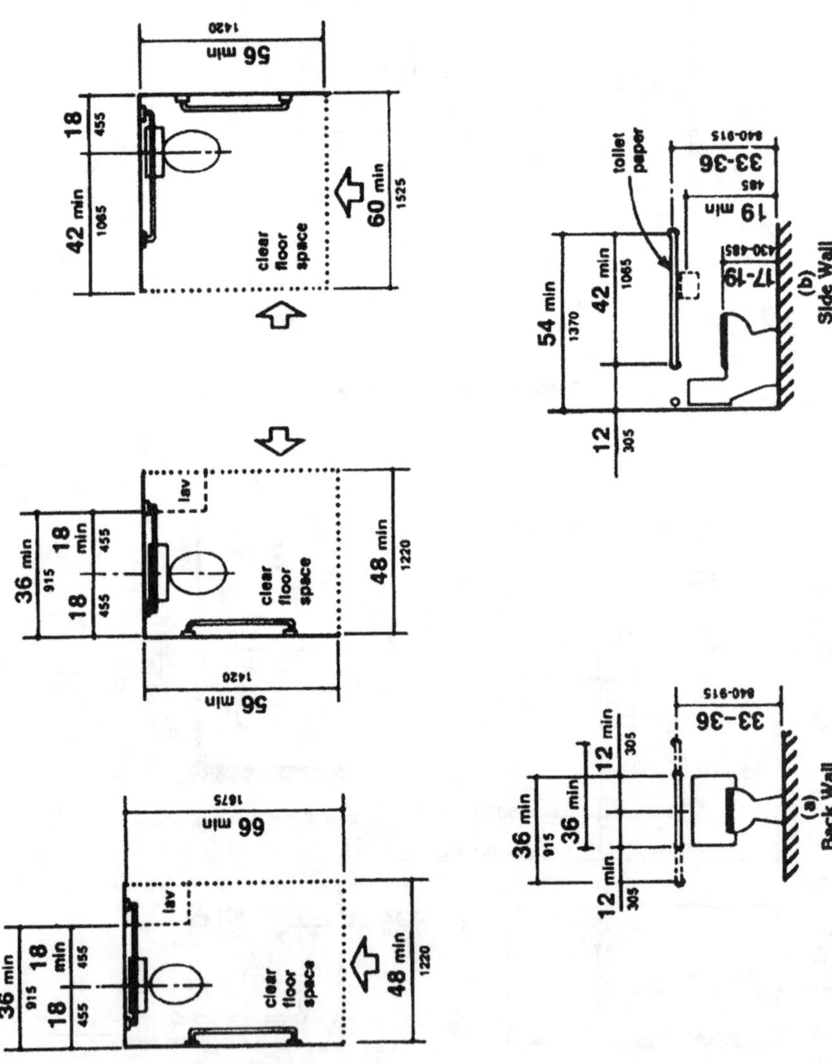

Figure 8-2. Toilet Stalls (Courtesy: United States Department of Justice)

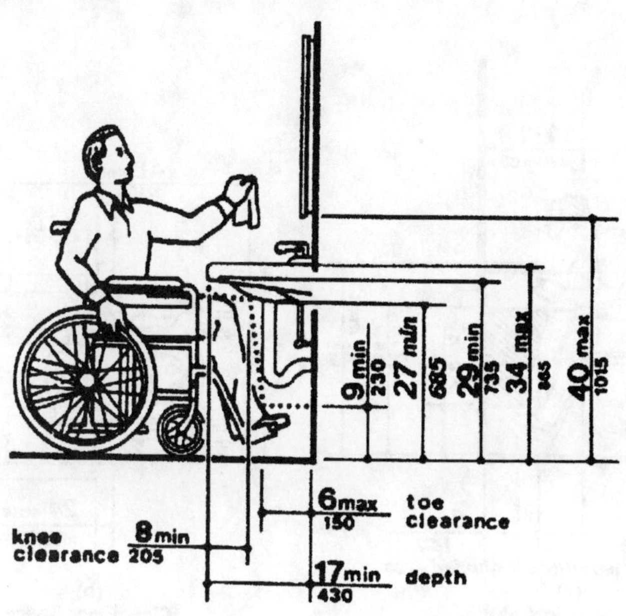


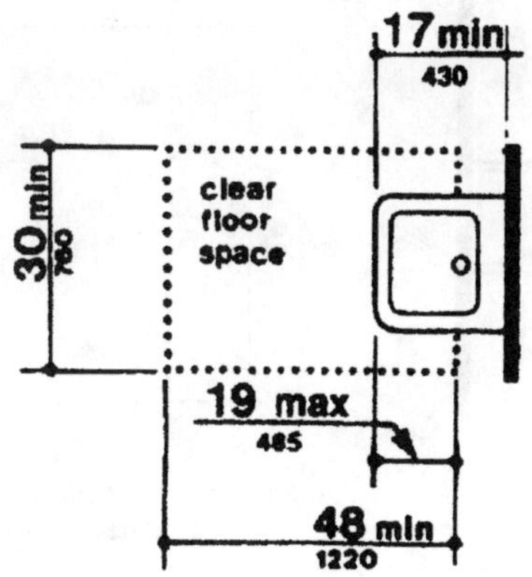


Figure 8-3. Lavatory Clearances
(Courtesy: United States Department of Justice)

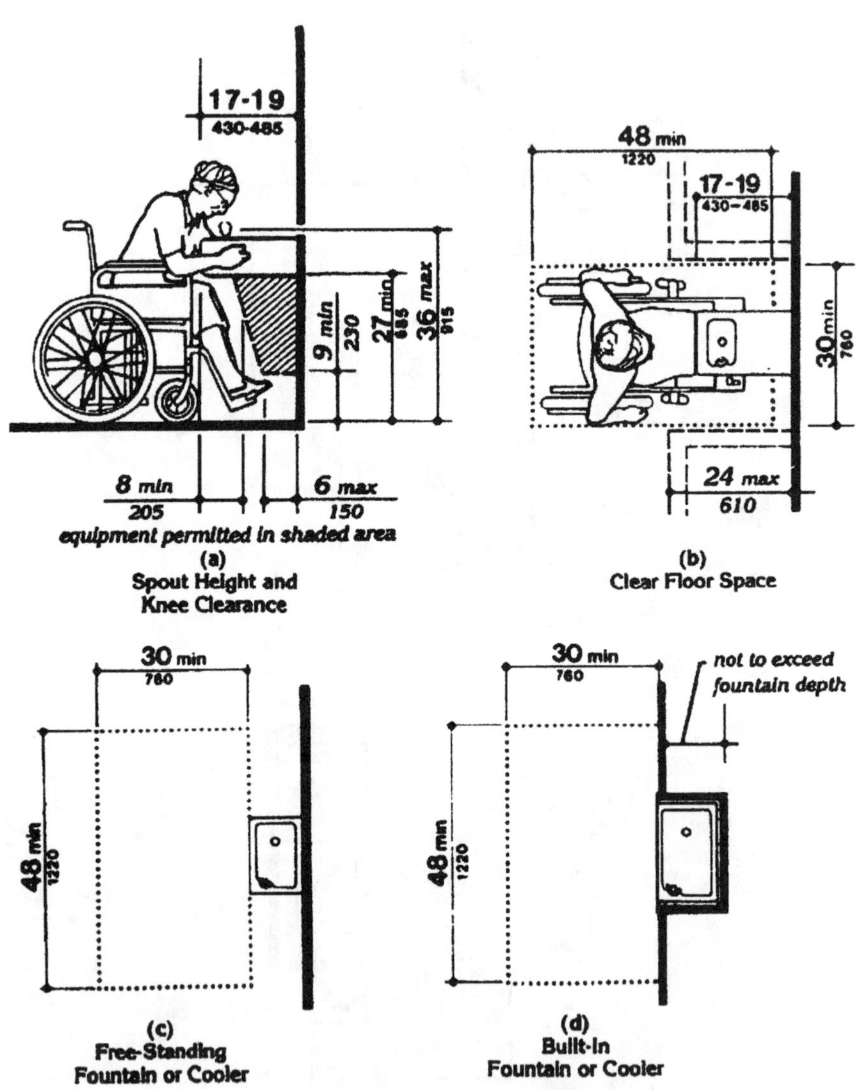


Figure 8-4. Drinking Fountains and Water Coolers
(Courtesy: United States Department of Justice)

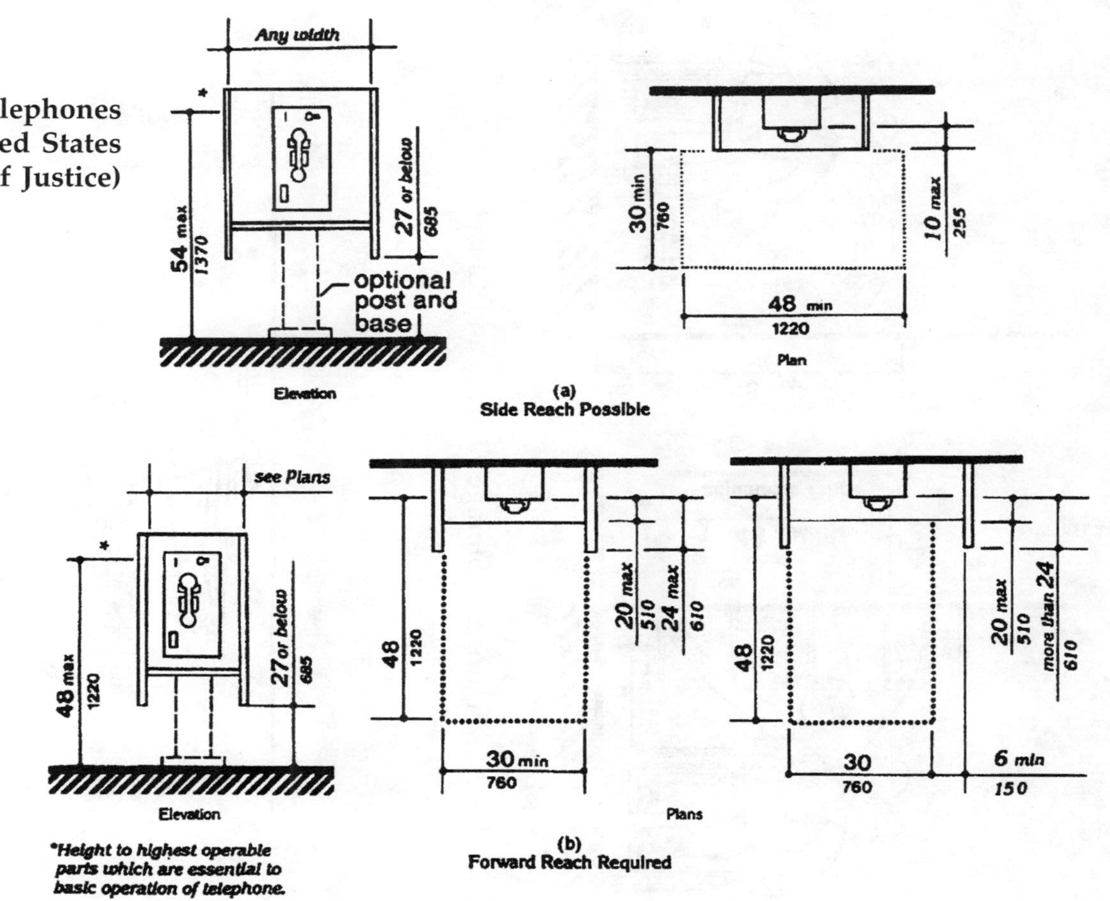


Figure 8-5. Telephones (Courtesy: United States Department of Justice)

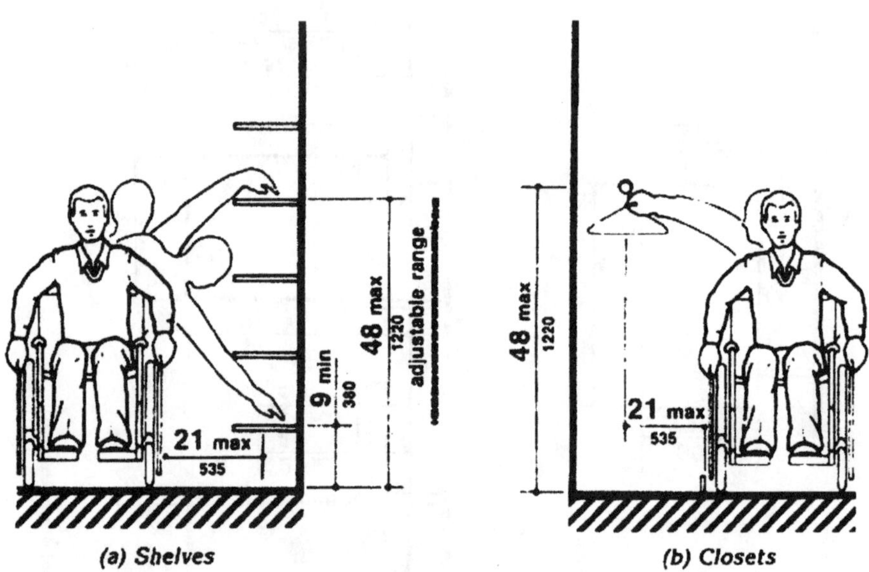


Figure 8-6. Storage Shelves and Closets

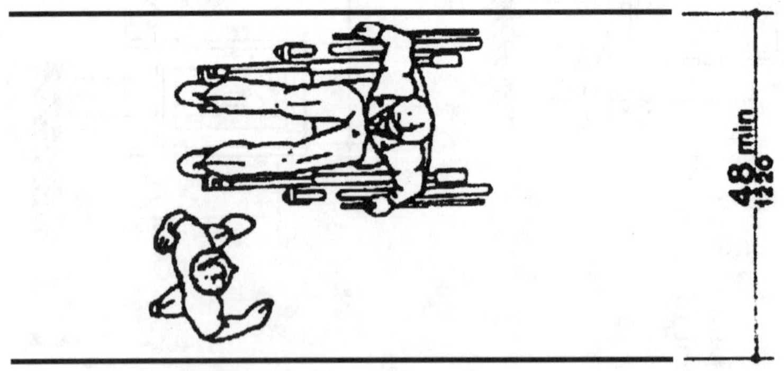


Figure 8-7. Wheelchair Clearance
(Courtesy: United States Department of Justice)

## Doors at Entrances to Businesses

Most entrances to stores and businesses use 36-inch-wide doors that are wide enough to be accessible. However, some older doors are less than 36 inches wide and may not provide enough width (32 inch clear width when fully opened). Door openings can sometimes be enlarged. It may also be possible to use special "swing clear" hinges that provide approximately 1-1/2 inches more clearance without replacing the door and door frame.

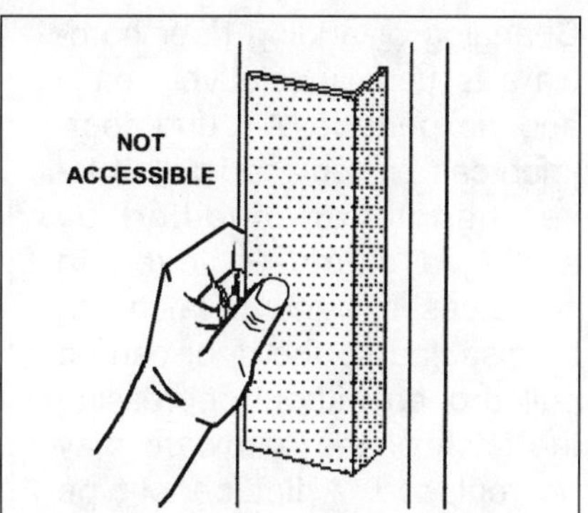


**This handle with a thumb latch is not accessible because one must grasp the handle and pinch down on the thumb latch at the same time.**

Inaccessible door hardware can also prevent access to the business. For example, the handle shown above requires the user to tightly grasp the handle to open the door. Many people with mobility disabilities and others with a disability that limits grasping, such as arthritis, find this type of handle difficult or impossible to use.

Other types of door hardware, such as a round door knob (which requires tight grasping and twisting to operate) or a handle with a thumb latch (at right) are also inaccessible and must be modified or replaced, if doing so is readily achievable.

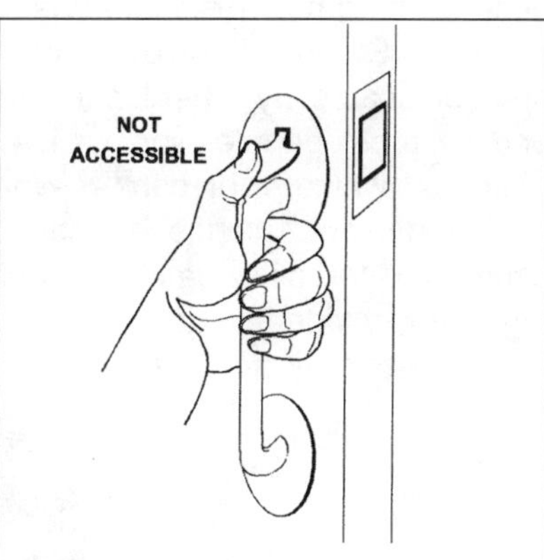


**This panel-type handle is not accessible because it requires the user to tightly grasp the handle to pull the door open.**

**Figure 8-8. Handles, Latches, Turnstiles
(Courtesy: United States Department of Justice)**

*(Continued)*

Changing or adding door hardware is usually relatively easy and inexpensive. A round doorknob can be replaced with a lever handle or modified by adding a clamp-on lever. In some cases, a thumb latch can be disabled so the door can be pulled open without depressing the latch or the hardware may be replaced. A flat panel-type pull handle can be replaced with a loop-type handle.

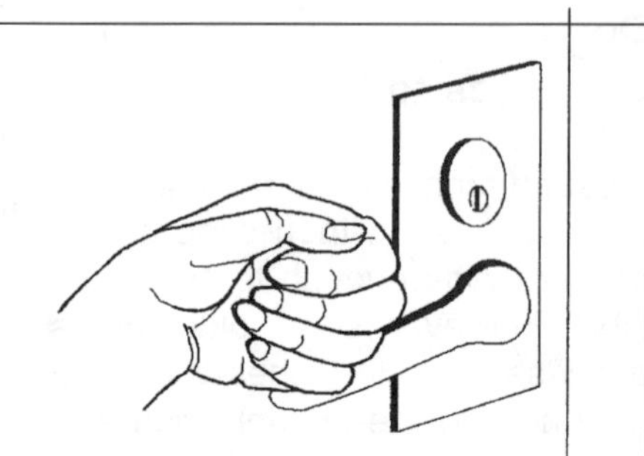

**A lever handle is accessible because it can be operated without tight grasping, pinching or twisting.**

## Turnstiles and Security Gates at Entrances

Businesses with narrow revolving turnstiles located at the entrance exclude people with disabilities unless accessible gates or passages are provided. Standard narrow turnstiles are not usable by wheelchair users and by most people who walk with crutches, walkers, or canes. Whenever a narrow turnstile is used, an accessible turnstile, gate or opening must be provided, if doing so is readily achievable.

**A loop-type handle is also accessible because it can be used without grasping, pinching or twisting.**

**Figure 8-8. Handles, Latches, Turnstiles** *(Concluded)*

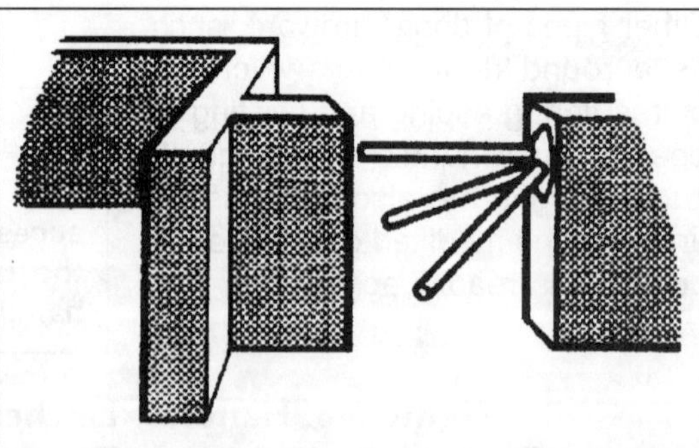

**This type of turnstile is not accessible to most people with disabilities.**

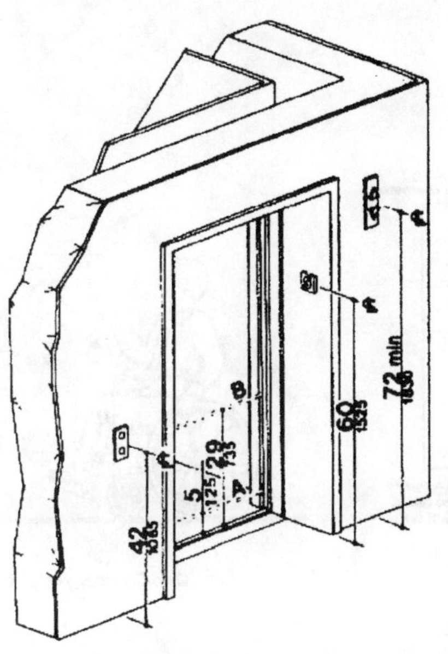

**Figure 8-9. Elevator Entrances**
**(Courtesy: United States Department of Justice)**

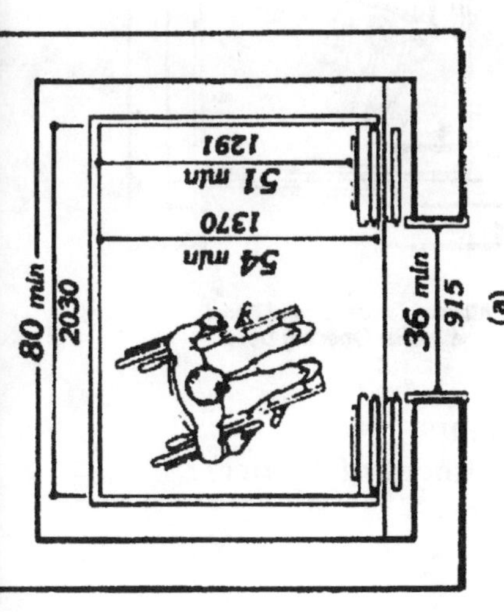

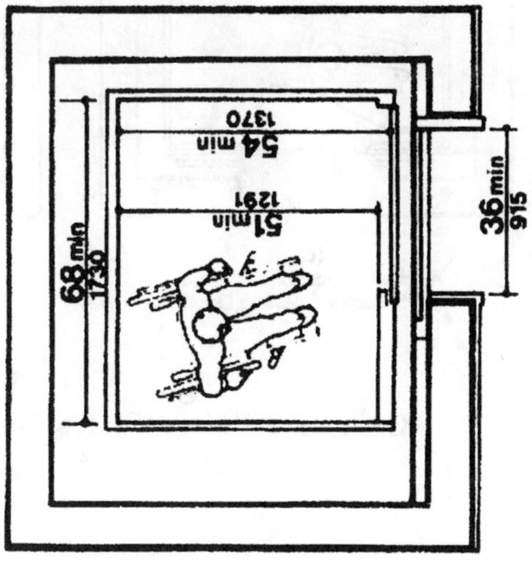

Figure 8-10. Elevator Cars
(Courtesy: United States Department of Justice)

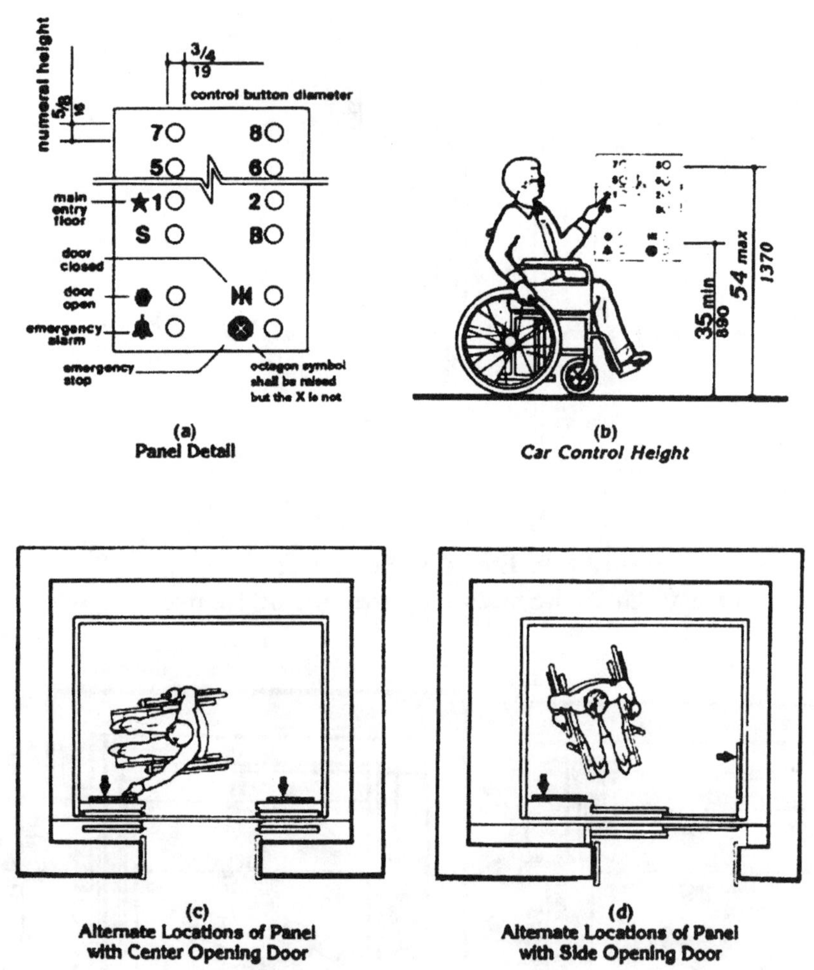


Figure 8-11. Elevators
(Courtesy: United States Department of Justice)

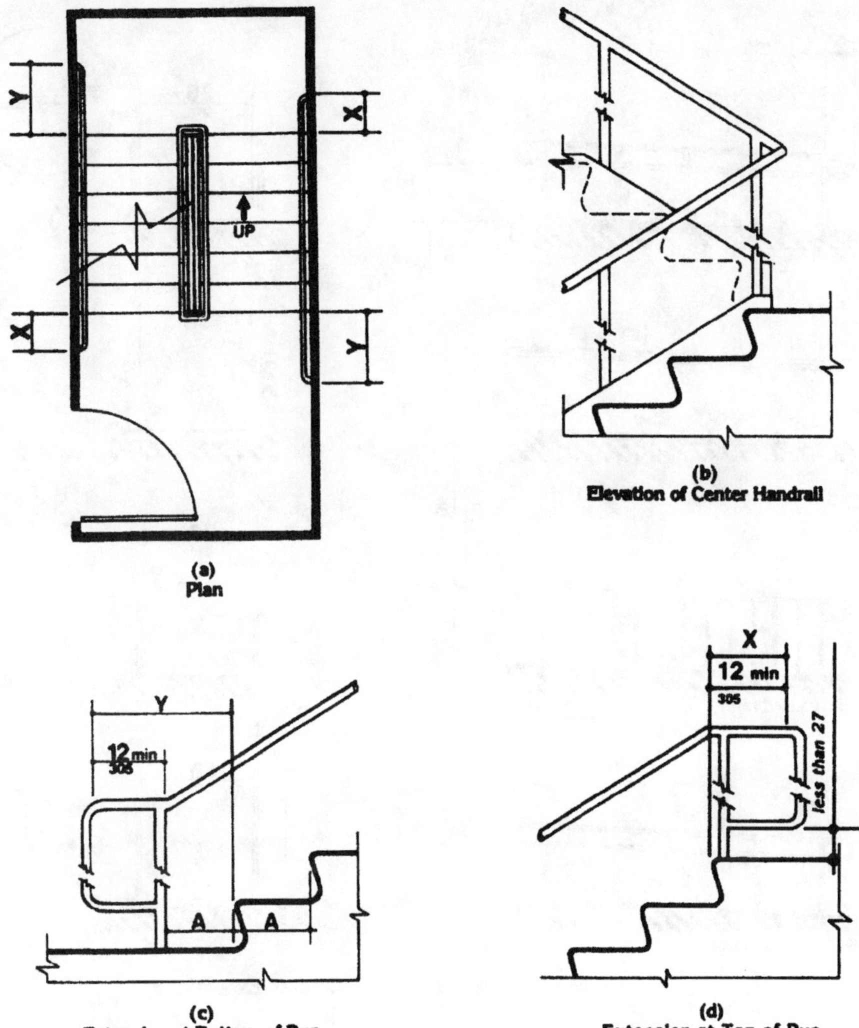



NOTE:

X   is the 12 in. minimum handrail extension
     required at each top riser.
Y   is the minimum handrail extension of 12 in.
     plus the width of one tread that is required
     at each bottom riser.

**Figure 8-12. Stair Handrails
(Courtesy: United States Department of Justice)**

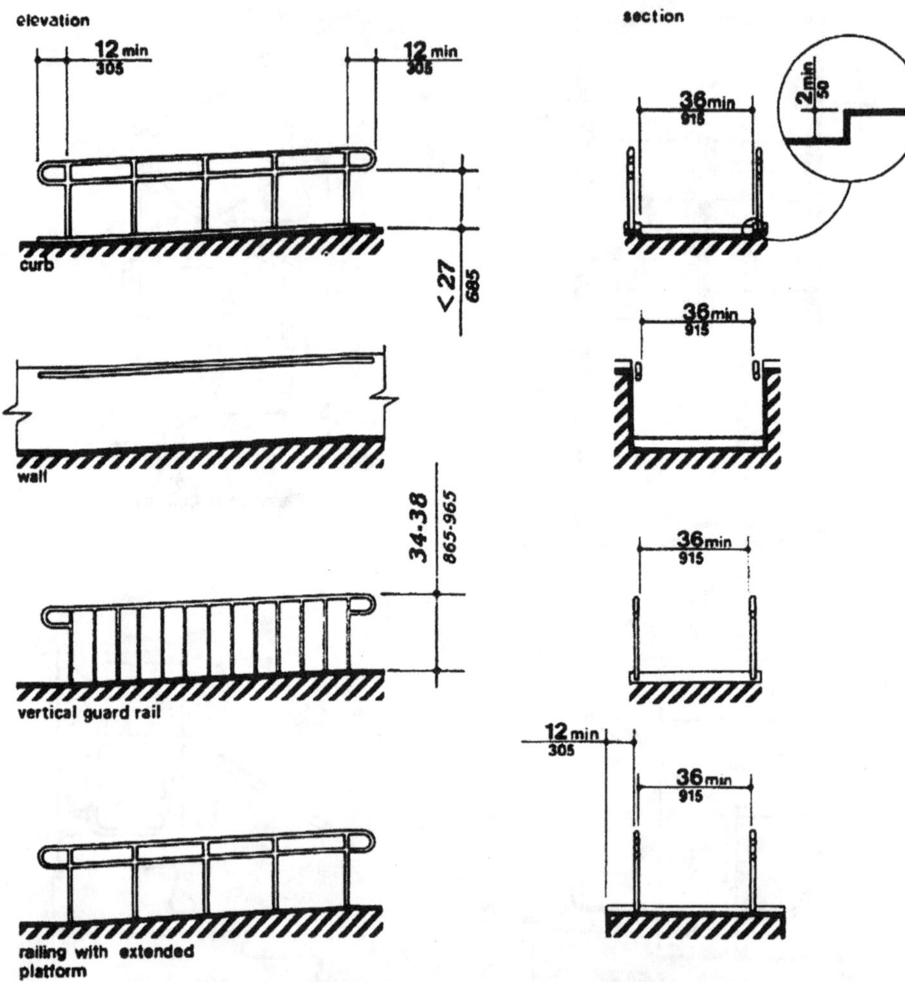



**Figure 8-13. Handrails (*Continued*)**
**(Courtesy: United States Department of Justice)**

A primary function area is any area within the facility where a major activity takes place. Customer service areas, work areas in places of public accommodation and all offices and work areas in commercial facilities are considered primary function areas.

Alterations to primary function areas must be done in compliance with ADAAG. Alterations must provide for and include an accessible path of travel from the altered area to the entrance.

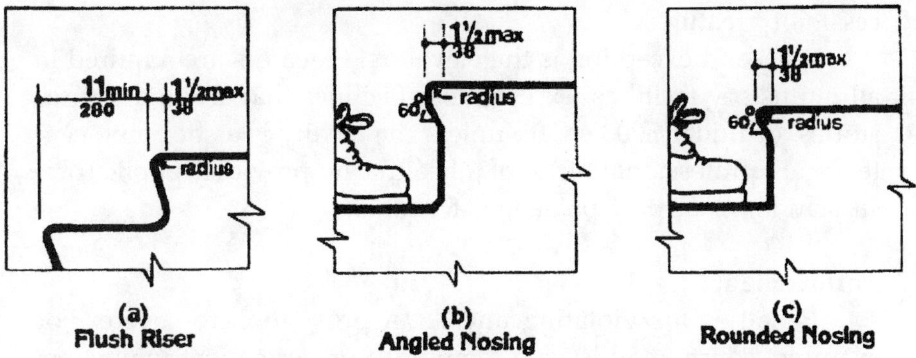


**Usable Tread Width and Examples of Acceptable Nosings**

**Figure 8-13. Handrails (*Concluded*)**

The path of travel must be a continuous route that connects the altered primary function area to the entrance, and must include the restrooms, telephones and drinking fountains that serve the area that is altered.

Any alteration made in compliance with the ADAAG must be made to the extent that is technically feasible to do so. If the alteration is not technically feasible, compliance is waived. Cost is not a consideration to determine if the alteration is technically feasible.

*New Construction*

Newly constructed facilities must be in strict compliance with ADAAG to ensure they are readily accessible to and usable by persons with disabilities.

Exceptions, however, are made. The standards do allow for those unusual circumstances when the unique characteristics of the land prevent incorporation of accessibility features in the facility. When this occurs, incorporating accessibility features in the new facility becomes structurally impracticable. The U.S. DOJ, which enforces Title III provisions, expects that this exception will be used rarely and in the most unusual of circumstances. Even when it is used, the remainder of the facility that can be constructed with accessibility features must be constructed with accessibility features.

A second exception is the elevators. Elevators are required in all multistory facilities, except those facilities that are under three stories, or under 3000 sq. ft., unless the facility is a shopping center/mall; professional office of a health care provider; public transit station; or airport passenger terminal.

**Enforcement**

Penalties for violating any ADA provision are severe. For example, courts may issue a temporary or permanent injunction, restraining order, or other order to an employer as a result of any civil action initiated by persons who have been subjected to discrimination, or believe that they are about to be subject to discrimination.

Civil penalties of up to $50,000 for a first violation and penalties up to $100,000 for subsequent violations may be imposed against a company found in violation of ADA provisions. While punitive damages are not available, the prevailing party may be awarded reasonable attorneys' fees, including litigation expenses and costs.

Additionally, a court may order the employer to alter its facilities so that they are accessible to and usable by people with disabilities; provide auxiliary aids and/or services; and modify any policy, practice or procedure.

Under the ADA, public accommodations carry specific obligations, including making reasonable modifications to policies, practices and procedures that are necessary to providing equal goods and services to people with disabilities. When a business owner/facility manager removes barriers, that business should follow the design requirements for new construction in the ADA Standards for Accessible Design (Standards). There are some cases that existing conditions, limited resources (in the case of a small business) or both will not make it not "readily achievable" to follow these standards fully. If this occurs, the ADA states that barrier removal measures may deviate from the Standards so long as the measures do not pose a significant risk to the health or safety of individuals with disabilities or others.

What are the priorities for barrier removal? The U.S. Department of Justice, Civil Rights Division, suggests that the first priority is to provide access to the business from public sidewalks, parking, and public transportation and then provide access to the areas where goods and services are made available to the public. Once these barriers are removed, access to public toilet rooms should be provided along with those barriers that limit use of public telephones and drinking fountains.

A common adjustment and/or modification that most businesses, companies, and/or facilities deal with is the accessible parking space. When a business restripes a parking lot, it must provide accessible parking spaces are required by the ADA Standards for Accessible Design.

In addition, businesses or privately owned facilities that provide goods or services to the public have a continuing ADA obligation to remove barriers to access in existing parking lots when it is readily achievable to do so. When parking is provided for the public, designated accessible parking spaces must be provided. Because restriping is relatively inexpensive, it is readily achievable in most cases.

## Accessible Parking Spaces for Cars

According to the ADA, accessible parking spaces for cars have at least a 60-inch-wide access aisle located adjacent to the designated parking space. The access aisle is just wide enough to permit a person using a wheelchair to enter or exit the car. These parking spaces are located on level ground and identified with a sign that is located in front of the parking space and mounted high enough so it is not hidden by a vehicle parked in the space. In addition the accessible parking space should be located closest to the building's accessible entrance. An accessible route must also be provided between the access aisle and the accessible building entrance. This route must have no steps or steeply sloped surfaces and it must have a firm, stable, slip-resistant surface. The recommendation is a 1:50 maximum slope in all directions if readily achievable.

**Van-Accessible Parking Spaces**

Van-accessible parking spaces are the same as accessible parking spaces for cars except for three features needed for vans:

- A wider access aisle (96") to accommodate a wheelchair lift;

- Vertical clearance to accommodate van height at the van parking space, the adjacent access aisle, and on the vehicular route to and from the van-accessible space, and

- An additional sign with the international symbol of accessibility that identifies the parking spaces as "van accessible." It should be high enough not to be hidden by the van parked in its space. And although the space is designated a van accessible space, cars may use the space too.

**Location**

The ADA further indicates where the accessible parking is to be located. According to the ADA, "Accessible parking spaces must be located on the shortest accessible route of travel to an accessible facility entrance. Where buildings have multiple accessible entrances with adjacent parking, the accessible parking spaces must be dispersed and located closest to the accessible entrances."

The ADA further indicates that when accessible parking spaces are added in an existing parking lot, locate the spaces on the most level ground close to the accessible entrance. An accessible route must always be provided from the accessible parking to the accessible entrance. An accessible route never has curbs or stairs, must be at least 3 feet wide, and has a firm, stable, slip-resistant surface. The slope along the accessible route should not be greater than 1:12 in the direction of travel.

Accessible parking spaces may be clustered on one or more lots if equivalent or greater accessibility is provided in terms of distance from the accessible entrance, parking fees, and convenience. Van-accessible parking spaces located in parking garages may be clustered on one floor (to accommodate the 98-inch minimum vertical height requirement).

(See Figure 8-14) A Checklist for the Parking and Drop-Off Areas (ADAAG 4.6).

Accordingly, any disaster and recovery plan must include provisions for people with disabilities, including employees and occupants, tenant-occupants, transient occupants and the general public.

| QUESTIONS | | POSSIBLE SOLUTIONS |
|---|---|---|
| **Parking and Drop-Off Areas (ADAAG 4.6)**<br>Are an adequate number of accessible parking spaces available (8 feet wide for car plus 5-foot access aisle)? For guidance in determining the appropriate number to designate, the table below gives the ADAAG requirements for new construction and alterations (for lots with more than 100 spaces, refer to ADAAG):<br><br>**Total spaces** **Accessible**<br>  1 to 25    1 space<br>  26 to 50    2 spaces<br>  51 to 75    3 spaces<br>  76 to 100    4 spaces<br><br>Are 8-foot-wide spaces, with minimum 8-foot-wide access aisles, and 98 inches of vertical clearance, available for lift-equipped vans?<br><br>**At least one of every 8 accessible spaces** must be van-accessible (with a minimum of one van-accessible space in all cases). | Yes  No<br><br>□  □<br><br>___<br>number of accessible spaces<br><br>Note widths of existing accessible spaces:<br><br>□  □<br><br>___<br>width/ vertical clearance | □ Reconfigure a reasonable number of spaces by repainting stripes.<br><br><br><br><br>□ Reconfigure to provide van-accessible space(s). |
| Are the access aisles part of the accessible route to the accessible entrances? | □  □ | □ Add curb ramps.<br>□ Reconstruct sidewalk. |
| Are the accessible spaces closest to the accessible entrance? | □  □ | □ Reconfigure spaces. |
| Are accessible spaces marked with the International Symbol of Accessibility? Are there signs reading "Van Accessible" at van spaces? | □  □ | □ Add signs, placed so that they are not obstructed by cars. |
| Is there an enforcement procedure to ensure that accessible parking is used only by those who need it? | □  □ | □ Implement a policy to check periodically for violators and report them to the proper authorities. |

Checklist for Existing Facilities version 2.1. © revised August 1995, Adaptive Environments Center, Inc. for the National Institute on Disability and Rehabilitation research. Reprinted with permission.

**Figure 8-14. "Checklist for the Parking and Drop-Off Areas (ADAAG 4.6)**

**Sources**

Gustin, Joseph F. *Safety Management: A Guide For Facility Managers*, New York: UpWord Publishing, Inc., 1996.

United States Department of Justice, *Code of Federal Regulations*, 28 CFR Part 36, Final Rule.

United States Department of Justice, Civil Rights Division Disability Rights Section, *Title II Highlights*.

United States Department of Justice, Civil Rights Division Disability Rights Section, *Title III Highlights*.

United States Department of Labor, Occupational Safety And Health Administration, "Hazardous Waste Operations And Emergency Response," Title 29, *Code of Federal Regulations*, 1910.120.

United States Department of Labor, Occupational Safety And Health Administration, *Principal Emergency Response And Preparedness Requirements in OSHA Standards And Guidance for Safety And Health Programs*, OSHA 3122, 1993 (Rev.).

United States Department of Labor, Occupational Safety And Health Administration, "Process Safety Management of Highly Hazardous Chemicals," Title 29, *Code of Federal Regulations, Part 1910.119.*

# Chapter 9

# The Safety, Emergency Response & Hazard Communication Planning Program

## PURPOSE

There are a number of reasons why a written safety and health plan is important.

First a well-written safety and health plan ensures compliance with the myriad of safety and health regulations mandated by OSHA and other agencies governing workplace safety and health.

Second, it provides a mechanism for incorporating other workplace mandates, such as the Americans with Disabilities Act (ADA) and worker's compensation statutes, into the company's overall hazard prevention program.

Third, it serves as the basis of the essential framework of a company's commitment to the shared safety responsibility of employers and employees.

Fourth, it allows for the equitable application and enforcement of policies, procedures, practices and rules. Equity is particularly relevant from a management perspective because it presumes consistency. Consistency in applying and enforcing rules, regulations, policies and procedures is what, in many cases, keeps managers out of trouble; and often times, consistency is what keeps employers out of court.

Finally, under OSHA's current penalty structure, a company with a written safety and health plan that demonstrates commitment and involvement by all personnel receives a 25% penalty reduction based upon its good faith efforts. A 25% reduction in penalties is reason enough for facility managers to develop a written safety and health plan, while making it an "easy sell" to their employers.


## ELEMENTS AND SCOPE OF THE WRITTEN SAFETY PLAN

In January 1989, OSHA issued guidelines for managing safety and health programming. These guidelines, while voluntary in nature, serve as the basis for developing a written safety and health management plan.

The guidelines are specific in identifying the elements that OSHA defines as essential to an effective safety and health management program. These elements are:

1. Management Commitment and Employee Involvement
2. Worksite Analysis
3. Hazard Prevention and Control
4. Training

The written plan should reflect the key components of effective safety and health programming that are defined by each of the four elements:

Management commitment and employee involvement are paramount to successfully implementing effective programming. OSHA, in fact, describes management commitment and employee involvement as complementary, with

1). management providing the motivation and resources of organizing and controlling activities within the organization and

2). employee involvement providing the means by which workers develop and/or express their own commitment to safety and health protection for themselves and their co-workers. In tandem, management commitment and employee involvement form the "core" of a company's occupational safety and health program, and provide facility managers with direction in compliance planning.

Management commitment and employee involvement include:

- a clearly stated worksite policy on safe and healthful work and working conditions, so that all personnel with site responsibility (and personnel at other locations with responsibility of the site) fully understand the priority and importance of safety and health protection in the organization;

- a clearly communicated goal for the safety and health program that defines objectives for meeting that goal, so that all members of the organization understand the results desired and the measures planned for achieving them;

- visible top management involvement in implementing the program, so that all employees understand that management's commitment is serious.

- provisions for employee involvement in the structure and operation of the program and in the decisions that affect their safety and health, so that they will commit their insight and energy to achieving the safety and health program's goal and objectives;

- clearly communicated and assigned responsibilities for all aspects of the program, so that managers, supervisors and employees in all parts of the organization know what performance is expected of them;

- providing adequate authority and resources to the responsible parties, so that assigned responsibilities can be met;

- holding managers, supervisors and employees accountable for meeting their responsibilities, so that essential tasks will be performed;

- reviewing program operations at least annually to evaluate their success in meeting the goals and objectives, so that deficiencies can be identified and the program and/or the objectives can be revised when they do not meet the goal of effective safety and health protection.

Worksite analysis involves various examinations of all worksite operations and job functions, so that any existing hazards and conditions can be identified. Worksite analysis includes:

- conducting comprehensive baseline worksite surveys for safety and health and periodic comprehensive update surveys;

- analyzing planned and new facility, processes, materials and equipment;

- conducting regular site safety and health inspections so that new or previously missed hazard and failures in hazard controls are identified;

- providing a reliable and reprisal-free system for employees to notify management personnel about conditions that appear hazardous and to receive timely and appropriate response (this utilizes employee insight and experience in safety and health protection and allows employee concerns to be addressed);

- investigating accidents and "near miss" incidents so that their causes can be identified and their prevention can be determined;

- analyzing injury and illness trends over time so that patterns with common causes can be identified and prevented.

**Hazard prevention and control** involves preventing the hazard when and where feasible, by effective job and/or worksite design and/or instituting control procedures to either eliminate or minimize hazard effects.

The procedures should address measures that:

- use engineering control techniques where feasible and appropriate;

- establish, at the earliest time, safe work practices and procedures that are clearly understood and followed by all affected parties (this includes a clearly communicated disciplinary system);

- use administrative controls, such as reducing the duration of exposure;

- provide personal protective equipment when necessary and/or when engineering and administrative controls are not feasible;

- maintain the facility and all equipment to prevent equipment breakdown;

- plan and prepare for emergencies by conducting training and emergency drills as needed to ensure the proper and safe responses to emergencies will be "second nature" for all persons involved, including the leaders who will be expected to manage and coordinate emergency response activities.

- develop a hazard communication program that alerts workers, as well as contractors and their employees to the potential risks; and

- establish a medical management program that includes on-site first aid, as well as nearby physician/emergency medical care to reduce the risk of any injury and/or illness that occurs.

**Training** that is job-specific must be provided to all employee levels so that they fully understand and are aware of the hazards which they may be exposed to while performing their job duties. Supervisors should be trained to:

- analyze work under their supervision to anticipate and identify potential hazards;

- maintain physical protections in their work areas;

- reinforce employee training on the nature of potential hazards in their work and on needed protective measures, through continual performance feedback and, if necessary, through enforcement of safe work practices; and

- understand their safety and health responsibilities.

**Multi-site Facilities**: While corporate offices normally develop safety and health policies, procedures and practices, a company's site facilities must also develop their own "site-specific" compliance plans. These site-specific plans must address the specific operations of the site. The site's designated safety officer assumes responsibility for compliance planning.


## GETTING STARTED

Prior to writing the actual plan, there are a number of preliminary steps that facility managers must take in order to assess their company's overall compliance with safety and health standards and regulations. These preliminary steps include:

- Reviewing the company's existing safety policy;

- Analyzing current worksite practices and procedures; and
- Conducting periodic self-inspection.

Each of these steps is critical to meeting OSHA's requirement for a "safe and healthful working environment" for all employees. While the first two steps should be considered as the essential prerequisites that form the framework of the written safety and health plan, the third step—conducting periodic self-inspections—provides the basis for updating the written plan, as updating or modifying the plan becomes necessary.

**Reviewing the Company's Existing Safety Policy**: Policies are the official governing principles of a company. As such, policies determine the course of action that a company takes in terms of its mission and methods of operation. In short, policies provide a company's direction.

**For this reason the safety policy's general statement should reflect:**

- The involvement of all company personnel in safety issues; and

- The commitment of all company personnel to maintaining a safe work environment.

A good example of a general policy statement that captures the essence and spirit of shared involvement and commitment to safety is provided by OSHA in the Agency's Handbook for Small Businesses:

*"We recognize that the responsibilities for safety and health are shared:*

- *The employer accepts the responsibility for leadership of the safety and health program, for its effectiveness and improvement, and for providing the safeguards required to ensure safe conditions.*

- *Supervisors are responsible for developing the proper attitudes to-*

*ward safety and health in themselves and in those they supervise, and for ensuring that all operations are performed with the utmost regard for the safety and health of all personnel involved, including themselves.*

- *Employees are responsible for wholehearted, genuine cooperation with all aspects of the safety and health program including compliance with all rules and regulations—and for continuously practicing safety while performing their duties.*

To ensure that all persons governed by the policy understand their rights and responsibilities, the policy must clearly communicate the company's position on the issue. The policy text, therefore, should include the safety goals that are company-specific, and the methods that the company will use to achieve those goals. This is accomplished by the process shown below:

1.  Define objectives for meeting the safety goals;

2.  Assign specific responsibilities for all employee levels so that all persons fully understand the consequences of noncompliance;

3.  Determine accountability for various employee levels so that all persons fully understand the consequences of noncompliance;

4.  Provide the authority, as appropriate, and the resources as necessary, to all employee levels so that they can meet their specific responsibilities; and

5.  Establish a mechanism for employees to report any hazards and/or potential hazards without fear of reprisal.

*Analyzing Worksite Practice and Procedures*
Assessing compliance involves analyzing the current conditions at a facility and involves two major activities;

1.  Undertaking a comprehensive safety and health survey of the entire facility; and

2.  Evaluating existing safety and health programming so that areas of strength and weakness can be identified.

The initial survey should focus on evaluating workplace conditions with respect to safety and health regulations and generally recognized safe work practices. With such a focus, two additional benefits are derived. First, employee work habits and practices can be directly observed. Second, safety and health issues and concerns can be discussed directly with employees. OSHA's *Handbook for Small Businesses* contains an extensive self-inspection checklist that can be used to conduct the initial survey.

The second major activity in assessing an existing safety and health program is to identify areas that are working well, and to identify those areas that need improvement. Included in this activity is reviewing the four major components of a total safety and health management program. These components are:

1.  Safety and Health Activities, which involves evaluating current/past activities as well as examining operations and practices, guidelines, policies and training program needs.

2.  Equipment, which entails making lists of the company's major equipment, principal operations and the locations of each. Particular attention should be given to inspection schedules, maintenance activities and plan and office layouts.

3.  Employee Capabilities, which involves reviewing the employment history of workers, includes hire/transfer dates, pervious and current experience and training. Particular attention should be given to newly-hired employees; employees who have handicaps and employees who have been determined to have disabilities, per ADA criteria.

4.  Accident and Injury/Illness History involves tracking

worker's compensation records, insurance payments first-aid cases and employee attendance records. Tracking these records can point out incidents of recurring injury/illness and point out safety areas that need to be shored-up.

The data obtained from this "fact-finding" provides very valuable information to employers. They can pinpoint any interruptions in normal operations, equipment and personnel downtime and product flaws.

Depending upon the results of this assessment, problems can be identified. Once problems are "isolated," solutions to correcting them can be determined.

*Conducting Periodic Self-Inspections*

The third step in assessing compliance involves periodic self-inspection. Regularly scheduled, a self-inspection can point out hazards and potential hazards that may have been missed in the initial survey. Self-inspections can also identify deficiencies in hazard control methods and procedures and, as previously noted, form the basis for modifying the written safety plan, as such modification, or updating, becomes necessary. Specifically, a self-inspection should include:

- Processing, Receiving, Shipping and Storage—equipment, job planning, layout, heights, floor loads, projection of materials, material handling and storage methods.

- Building and Grounds Conditions—floors, walls, ceilings, exits, stairs, walkways, ramps, platforms, driveways, aisles.

- Housekeeping Program—waste disposal, tools, objects, materials, leakage and spillage, cleaning methods, schedules, work areas, remote areas, and storage areas.

- Electricity—equipment, switches, breakers, fuses, switchboxes, junctions, special fixtures, circuits, insulation, extensions, tools, motors, grounding, NEC compliance.

- Lighting—type, intensity, controls, conditions, diffusion, location, glare and shadow control.

- Heating and Ventilation—type, effectiveness, temperature, humidity, controls, natural and artificial ventilation and exhausting.

- Machinery—points of operation, flywheels, gears, shafts, pulleys, key ways, belts, couplings, sprockets, chains, frames, controls, lighting for tools and equipment, brakes, exhausting, feeding, oiling, adjusting, maintenance, lock-out, grounding, work space, location, and purchasing standards.

- Personnel—training, experience, methods, of checking machines before use, type of clothing, personal protective equipment, use of guards, tool storage, work practices, method of cleaning, oiling, or adjusting machinery.

- Hand and Power tools—purchasing standards, inspection, storage, repair, types, maintenance, grounding, use and handling.

- Chemicals—storage, handling, transportation, spills, disposals, amounts used, toxicity or other harmful effects, warning signs, supervision, training, protective clothing and equipment.

- Fire Prevention—extinguishers, alarms, sprinklers, smoking rules, exits, personnel assigned, separation of flammable materials and dangerous operations, explosion-proof fixtures in hazardous locations, waste disposal.

- Maintenance—regularity, effectiveness, training of personnel, materials and equipment used, records maintained, method of locking out machinery, general methods.

- Personal Protective Equipment—type, size, maintenance, repair, storage, assignment of responsibility, purchasing methods, standards observed, training in care and use, rules of use, method of assignment.

Easily adaptable to any company's particular situation, this outline can serve as the basis for the periodic self-inspection.

In a word, assessment means analysis. And while analyzing the worksite to ensure compliance with regulatory mandates might be its most obvious purpose, ensuring worker safety is its most immediate and compelling reason for doing so.

## WRITING THE SAFETY AND HEALTH PLAN: HOW TO DO IT

The written safety and health plan does not have to be lengthy; nor does it have to be complicated. As long as the essential elements are addressed, the written safety and health plan will meet its stated purpose.

As previously discussed, the four basic elements of the effective safety and health plan, deemed essential by OSHA, include:

- Management commitment and employee involvement;
- Worksite analysis;
- Hazard prevention and control; and
- Training.

Appropriately defined, each of these essential elements speaks to a facility's total safety and health program, including emergency preparedness and hazard communications. Both of these issues must also be included within the context of the company's overall safety and health program, either as sub-sections of the plan, or as separate, "stand-alone" documents.

**Review Standards**: In writing the safety and health plan, it is essential to first review all applicable standards and regulations, whether promulgated by OSHA or any other agency that governs the operations of a particular industry and/or the activities of a particular facility. In reviewing the standards and regulations, facility managers must consider both general industry standards—which have an almost "universal" application—and the standards which are specific to their particular industry.

The "key" factors to identify in the standards review include:

1.  Who is covered by the standard;

2.  What the standard requires in terms of controls, processes and methods of operation;

3.  Conditions under which those controls, processes and methods of operation are to be instituted and/or initiated; and

4.  Required training for those employees whose jobs involve performing duties that are governed by the mandated controls, processes and methods of operation.

Each of these factors is critical to an effective safety and health plan because each forms the basis for compliance.

**Develop Policy**: Following the review of all applicable standards, the next step is to develop the safety policy. The general policy statement previously described, can serve as the basis for developing a company-specific policy that defines:

*   Rights;
*   Responsibilities; and
*   Measures that the company will take to ensure workplace safety.

A further delineation of the rights and responsibilities of all employees, as well as the measures that the company will take to ensure workplace safety, can be broken down even further.

For example, most companies and entities already have a policy statement(s) that outlines proscribed rules of employee conduct and disciplinary procedures. The "Rules of Conduct/Discipline" normally contain:

A.  Specific policy statement.
B.  Applicability.
C.  Guidelines for policy administration.
D.  Classification of misconduct (i.e., minor vs. serious offenses).
E.  Classification of disciplinary action.
F.  List of rules (of conduct)-examples of misconduct.
G.  Retention of disciplinary actions.

Since this particular policy statement already address these issues, it should be included in the plan itself, or at least referenced.

**Flexibility**: Since a safety and health program must be flexible enough to accommodate the myriad of changes that occur in the workplace, its safety and health plan must also be flexible enough to reflect those changes. As a "working" document, a company's written plan must be flexible enough to accommodate the changes to its safety and health program. The change factors that influence an entity's program may occur as a result of new standards being introduced, revisions to existing standards, as well as the host of changes that occur naturally within a business setting.

In developing its essential elements for the effective safety and health program, OSHA recognized the need for flexibility. These elements provide the general direction that companies should take in developing their programs, yet provide companies with the latitude to determine how each element will be addressed.

## PUTTING THE PLAN TOGETHER

As previously noted, the written safety and health plan does not have to be lengthy, nor does it have to be complicated. However, consideration must be given to those "core" components that are integral to effective safety and health programming. As such, these "core" components must also be addressed within the written plan, itself. These components are:

**Regulatory standards review** in which all applicable industry mandates, including emergency action plans, are reviewed to ensure compliance with defined safe work processes, practices, procedures and required employee training in same.

**Policy review and development** in which a review of a company/facility's existing policies is conducted to determine if the issues of safety and health are appropriately addressed within the context of business operations. Included in the policy review, is the applicability of a company/facility's "Rules of Conduct/Discipline Policy." This specific policy must be incorporated into the overall safety and health plan, so that all employees are aware of the company/facility's serious commitment to safety. Including this policy also ensure that employees at all levels will understand that they will be held accountable for meeting their safety responsibilities.

**Regular worksite inspections** provide facility managers with the information needed to identify, as well as isolate any problems that are either occurring, or that may occur in the workplace. Routinely conducted, worksite inspections can identify any processes, practices and procedures that carry the potential for accident and/or injury. Routinely inspecting machinery, equipment and personal protective equipment can also help facility managers in identifying potential and/or existing hazards that may have been either previously missed, or which may have surfaced since previous inspections.

**Documentation**, which entails all aspects of a company/facility's operations, including safety and health programming, is critical for several reasons. First, effective documentation of all safety and health initiatives provides employers with the means to track the company's safety performance and to modify and/or adapt operations, processes, procedures and training when and where necessary to meet safety/health goals and objectives. Second, effective documentation of all safety/health initiatives, strategies and activities, demonstrates to enforcement agencies that the company/facility takes its commitment to safety and health, seriously.

**Training** serves as the means for providing employees

with the basis for performing their assigned job tasks in a safe manner. On-going training in work processes, practices and procedures as well as in basic emergency preparedness procedures enables employees to safely meet their responsibilities, while at the same time reinforcing the fact that they will be held accountable for the safe performance of their job duties.

**Sources**

Gustin, Joseph F., *The Safety, Emergency Response & Hazard Communication Planning Program*, New York: UpWord Publishing, 1996.

# Chapter 10
# Bomb Threats

T he Bureau of Alcohol, Tobacco And Firearms reported that from 1990—1994, 306 people were killed and more than 3,400 were injured as a result of explosives incidents. Total reported property damage exceeded $1.1 billion. Bombing and the threat of being bombed are realities that exist in today's world. Today, corporations and other organization must face the prospect that this type of violence could occur at their facilities.

## THE NATURE OF BOMBS

Bombs can be constructed to look like most any other object and they can be placed or delivered in a number of ways. According to the ATF, the probability of finding a bomb that looks like the "typical" bomb is almost nonexistent. Bombs and incendiary devices can take any shape or form. And whatever form or shape that they do take, they all have one singular purpose—they are intended to explode. Most bombs are home-made. Their design is limited by only two factors: the bomber's imagination and the resources that are available to the bomber.

## BOMB THREATS

Bomb threats are delivered in a variety of ways. In most cases, the majority of the threats are called in to the target. On occasion, bomb threats are called in through a third party, or are communicated to the target in writing or by a recording that is sent to the target. There are two logical explanations for reporting a bomb threat.

First, the caller has definite knowledge, or believes that an explosive or incendiary bomb has been placed at the target site and wants to minimize personal injury or property damage. The caller may be either the person who placed the explosive, or somebody who has knowledge that a device has been placed.

Second, the caller wants to create an atmosphere of anxiety as well as panic which, in all likelihood, will disrupt the normal activities at the target site.

Regardless of the motive, though, there will be reactions to the threat. Through effective planning, however, reactions can be controlled and an orderly response to the threat can be accomplished.

**Planning For the Bomb Threat**

Thorough and comprehensive planning accomplishes several primary objectives. The effective plan:

- Reduces the accessibility of a business or building.

- Identifies those areas within the facility that can be "hardened" against the potential bomber.

- Limits the amount of time lost to conduct a search, if such a determination is made.

- Instills confidence in the leadership of the management team, by reinforcing occupants' trust and belief that the persons in charge do care and that every precaution is being taken to ensure their personal safety.

Finally, thorough and comprehensive planning reduces the potential for panic which, in the context of the bomb threat, is the ultimate goal of the caller. Panic that is excessive—unreasoning terror—is the most contagious of all human emotions. Once a state of panic has been reached, the potential for personal injury and property damage is increased.

To effectively prepare for a bomb incident, it is necessary to develop two separate, but independent plans—the physical security plan and the bomb incident plan.

## The Physical Security Plan

The physical security plan provides for the protection of property, occupants, facilities and materials against the unauthorized entry, trespass, damage, sabotage or other illegal or criminal acts. In essence, the physical security plans deals with the prevention and control of access to the facility regardless of any existing security, which may not be designed or equipped to prevent a bomb attack incident.

## The Bomb Incident Plan

The bomb incident plan provides the detailed procedures that are to be implemented when a bombing attack is either threatened or actually implemented. In planning for such an occurrence, the bomb incident plan must establish a defined chain of command or line of authority, so that the incident will be handled with the least risk possible. The chain of command also helps to reinforce confidence in management, while minimizing the risk of panic.

### Lines of Authority: The Chain of Command

Establishing the lines of authority in the bomb incident plan is complex in both the single occupant organization and in the multi-occupant facility. In such cases, representatives from each floor (in the single occupant organization) and from each tenant-occupant (in the multi-occupant facility) should be involved in the planning process, with one person designated as the team leader/ coordinator. All people who serve on the planning team should have a clear understanding of their roles, their direct reporting relationship and the duties they are to perform in the event of a bomb incident.

Planning considerations should also include a designated command center. With a location in either the switchboard room or other focal point of telephone and/or radio communications, the command center serves as the hub of activities during the actual bomb incident. Staffed by the team leader/coordinator and designated members of the team who have the authority to make and execute the necessary decisions, a current, up- to-date floor plan or blueprint should be maintained in the command center. Lines of communication between the command center and the internal search and evacuation units are of paramount importance. The command center must have the flexibility to maintain ongoing contact with these units and to track their efforts.

**Government/Community Resources**—in developing the bomb incident plan, as well as the physical security plan, assistance from the local police and fire departments may be available. Additionally, local government agencies and the local offices of the ATF are available to provide assistance in developing effective strategies.

When possible, representatives from these organizations should be invited to the facility so that an in-depth survey of the building may be conducted. Such an on-site survey enables the planning team to pinpoint possible locations where a bomb could be placed. All possible locations where a bomb could be placed should be listed on a bomb checklist and retained in the command center. If bomb disposal units are provided by local enforcement agencies, decisions/agreements regarding the availability of those units in any search activities can be determined at that time.

A crucial element of any bomb incident or physical security plan is training. All personnel, including command center personnel and particularly the individuals who are assigned to the switchboard, must be thoroughly trained in all aspects of responding to a bomb threat (see Figure 10-1).

## SECURITY AGAINST BOMB INCIDENTS

The ATF has outlined various security measures that apply to "hardening" against the bomb attack. These security measures are discussed below.

## ATF Bomb Threat Checklist

Exact time of call ___________________________________________

Exact words of caller ________________________________________

_____________________________________________________________

_____________________________________________________________

QUESTIONS TO ASK

1. When is to bomb going to explode? _______________________
2. Where is the bomb? ______________________________________
3. What does it look like? __________________________________
4. What kind of bomb is it? ________________________________
5. What will cause it to explode? ___________________________
6. Did you place the bomb? ________________________________
7. Why? ___________________________________________________
8. Where are you calling from? _____________________________
9. What is your address? ___________________________________
10. What is your name? ______________________________________

CALLER'S VOICE (circle)

| | | | | |
|---|---|---|---|---|
| Calm | Disguised | Nasal | Angry | Broken |
| Stutter | Slow | Sincere | Lisp | Rapid |
| Giggling | Deep | Crying | Squeaky | Excited |
| Stressed | Accent | Loud | Slurred | Normal |

If voice is familiar, whom did it sound like? ________________

Were there any background noises? __________________________

Remarks: __________________________________________________

_____________________________________________________________

Person receiving call: ______________________________________

Telephone number call received at: _________________________

Date: ______________________________________________________

Report call immediately to: ________________________________

(Refer to bomb incident plan.)

**Figure 10-1. ATF bomb threat checklist. Courtesy: U.S. Department of The Treasury, Bureau of Alcohol, Tobacco And Firearms.**

Most commercial structures (as well as individual residences) have some form of security in place, whether planned or unplanned and whether realized or not. For example, locks on windows and doors, exterior lighting, etc., are all designed and installed to contribute toward the security of a facility and the protection of it occupants.

In considering measures to increase a facility's security, the local police department should be contacted for assistance. Their guidance is invaluable regarding any security enhancements that may be under consideration. Additional measures that should be addressed and which may reduce a building's vulnerability to bomb attacks include the following recommendations.

## Building Design

The exterior configuration of a building is, obviously, very important. While, in many instances, its architectural design does not consider the security precautions that must be taken in thwarting bomb attacks, a building's exposure to threat can be minimized by installing additional lighting, fencing and by controlling access.

Bombs that are either delivered by car or which are left in a car are a grave reality. Parking should be restricted, if possible, to 300 feet from the facility. In those situations where restricted parking is not feasible, all employee vehicles should be properly identified and parked closest to the facility itself. Parking for non-employee/occupants should be provided at a distance from the facility.

## Landscaping

Landscaping, particularly heavy shrubs and vines, should be maintained close to the ground to minimize either the perpetrator's cover, or "protection" for the bomb. Any ornamental or real window planters/boxes should be removed, unless there is an absolutely essential reason for their presence. Planters and window boxes are the "perfect" hiding places for explosives.

## Security Patrols

A highly visible security patrol can be a significant deterrent. Even if security consists of one person, that guard is optimally utilized outside the building. If the guard's services are utilized within the building, the installation of closed-circuit monitoring systems that cover exterior perimeters should be considered.

Burglar alarm systems that are adequate for the building's square foot dimensions are effective deterrents. These systems should be installed by reputable companies that can service and properly maintain the alarm equipment. Signs clearly warning that an alarm system is in place should be posted in very visible locations.

## Entrance and Exit Doors

Only entrance and exit doors with interior hinges and hinge pins should be installed. Solid wood or sheet metal-faced doors, as opposed to hollow-core doors, provide additional protection. A steel doorframe that properly fits the door is as important as the construction of the door itself.

## Window Protection

While the ideal security situation is the building with no windows, there are devices that can be installed which offer added measures of security. These devices include bars, gates, heavy mesh screens and steel window shutters. However, in considering any such enhancement, all local ordinances, including fire safety code, must be researched to ensure appropriateness.

## Access Control

Controls should be established for identifying personnel who are authorized access to critical areas within the facility and for denying access to unauthorized persons. These controls should extend to the inspection of all packages and materials that are being taken or "designated" for routing to critical building areas of "key" personnel.

Security and maintenance personnel should be alert for persons who act suspiciously. If any suspicious activity is detected, surveillance should begin. Also, security and maintenance personnel should be instructed to note any and all objects, items or packages that look "out-of-place" (Figure 10-2). Such suspicious items should be reported to the appropriate person/authority level for follow up. Any area within a facility where unusual activity has been noted should be inspected, especially potential hiding places for unauthorized personnel such as stairwells, restrooms and unoccupied/empty office spaces.

Doors and other access ways to such areas as computer rooms, boiler rooms, mail rooms, switchboards and elevator control rooms should remain locked when not in use. Procedures for key distribution and retention should be established and maintained by all facilities, regardless of occupancy type or size.

Environmental services, including housekeeping, are another major issue. Trash or dumpster areas should be free of debris and other contents. Bombs can easily be concealed in these areas. All combustible materials should be properly disposed of or protected if further use is anticipated.

Where feasible, detection devices should be installed. Placed at all entrances, as well as in those areas previously identified as likely places where a bomb may be placed, these detection devices serve as effective deterrents.

Heightened security measures should also be instituted in many of the buildings that are open to the public. Despite the minimal inconvenience to the public that "signing-in" and "signing-out" would incur, this kind of policy would add an extra measure of safety to all building occupants.

## RESPONDING TO BOMB THREATS

All building personnel, and particularly switchboard personnel, should be trained in bomb threat procedures. The ATF has developed guidelines to follow when a bomb threat is called into a facility. These guidelines are:

- It is always desirable that more than one person listens in on the call. To do this, a covert signaling system should be

## Department of the Treasury
## Bureau of Alcohol, Tobacco and Firearms
## Suspect Package Alert

- Is addressee familiar with name and address of sender?

- Package/letter has no return address.

- Is addressee expecting package/letter? If so, verify expected contents.

- Improper or incorrect title, address, or spelling of name of addressee.

- Title but no names.

- Wrong title with name.

- Handwritten or poorly typed addresses.

- Misspellings of common words.

- Return address and postmark are not from same area.

- Stamps (sometimes excessive postage, unusual stamps) versus metered mail.

- Special handling instructions on package (i.e., Special Delivery, Open by Addressee Only, Foreign Mail, Air Mail, etc.).

- Restrictive markings such as Confidential, Personal, etc.

- Overwrapped, excessive securing material such as masking tape, string, or wrappings.

- Lumpy or rigid envelopes (stiffer than normal, heavier than normal, etc.)

- Lopsided or uneven envelope.

- Oily stains or discolorations.

- Strange odors.

- Protruding wires or tinfoil.

- Visual distractions (drawings, unusual statements, hand-drawn postage, etc.).

*(Please be advised that this is only a general checklist. The best protection is personal contact with the sender of the package/letter.)*

FOR INFORMATION ON BOMB SECURITY OR BOMB THREATS, CONTACT YOUR LOCAL ATF OFFICE.

**Figure 10-2. ATF suspect package alert. Courtesy: U.S. Department of The Treasury, Bureau of Alcohol, Tobacco And Firearms.**

implemented, perhaps by using a coded buzzer signal to a second reception point.

- A calm response to the bomb threat caller could result in obtaining additional information. This is especially true if the caller wishes to avoid injuries or deaths. If told that the building is occupied or cannot be evacuated in time, the bomber may be willing to give more specific information on the bomb's location, components, or method of initiation.

- The bomb threat caller is the best source of information about the bomb. When a bomb threat is called in:

1. Keep the caller on the line as long as possible. Ask him/her to repeat the message. Record every word spoken by the person.

2. If the caller does not indicate the location of the bomb or the time of possible detonation, ask him/her for this information.

3. Inform the caller that the building is occupied and the detonation of a bomb could result in death or serious injury to many innocent people.

4. Pay particular attention to background noises such as motor running, music playing and any other noise which may give a clue as to the location of the caller.

5. Listen closely to the voice (male or female), voice quality (calm, excited), accents and speech impediments. Immediately after the caller hangs up, report the threat to the person designated by management to receive such information.

6. Report the information immediately to the police department, fire department, ATF, FBI and other appropriate agencies. The sequence of notification should be established in the bomb incident plan.

7.  Remain available as law enforcement personnel will want to conduct an interview.

When a written threat is received, save all materials, including any envelope or container. Once the message is recognized as a bomb threat, further unnecessary handling should be avoided. Every possible effort must be made to retain evidence such as fingerprints, handwriting or typewriting, paper and postal marks. This evidence proves essential in tracing the threat and in identifying the writer.

While written messages are usually associated with generalized extortion attempts, a written warning of a specific device may occasionally be received. When it is, it should never be ignored.

## EVACUATION AND SEARCH

Once a bomb threat has been called in or a written threat has been received, decisions have to be made. Essentially, there are three available options. The threat could simply be ignored; a decision to evacuate immediately could be made; or a building search could be undertaken, followed by an evacuation. If a decision to search then evacuate is made, a designated search unit is deployed.

### Search Units

Once the decision to search is made, more than one person should be used to conduct the search, no matter how small the room or area may be. A search unit can be comprised of supervisory personnel, area occupants, or trained explosive search teams. There are advantages and disadvantages to each method of staffing search units.

Using supervisory personnel to search is a rapid approach and causes little disturbance. There will be little loss of employee working time, but a morale problem may develop if it is learned that a bomb threat has been received and employees were not notified. Using supervisors to search will usually not be as thorough because of their unfamiliarity with many areas of the facility and their desire to return to work as soon as possible.

Using occupants to search their own areas is the best method for a rapid search. The occupants' concern for their own safety will contribute toward a more thorough search. Furthermore, the personnel conducting the search are familiar with what does or does not belong in their particular areas. Using occupants to search results in shorter loss of work time than if all persons were evacuated prior to the search by trained teams. Using the occupants to search also has a positive effect on morale, given a good training program to develop confidence. Obviously, the use of occupants requires the training of all building occupants and would ideally include performing several simulated training exercises. One drawback of the "occupant-search" method is the increased danger to unevacuated workers.

The search conducted by a trained team is the best for safety, morale and thoroughness, though it does take the most time. Using a trained team results in a significant loss of production time. It is a slow operation that requires comprehensive training and practice.

Any decision regarding the use of internal (i.e., supervisors/ occupants) versus external (i.e., professionally trained personnel) is, of course, a decision of the Disaster and Recovery Planning Team. Determination must be based upon company/building policy with occupant safety as the key consideration.

**Search Techniques**

While there are many possible variations to room searching, the ATF has outlined the basic techniques for searching a room. These techniques utilize a two-person search unit.

When the two-person search unit enters the room to be searched, they should first move to various parts of the room and stand quietly with their eyes closed and listen for a clockwork device. Frequently, a clockwork mechanism can be quickly detected without use of any special equipment. Even if no clockwork mechanism is detected, the team is now aware of the background noise level within the room itself.

Background noise or transferred sound is always disturbing during a building search. If a ticking sound is heard but cannot be located, one might become unnerved. The ticking sound may come from an unbalanced air-conditioner fan several floors away, or from a dripping sink down the hall. Sound will transfer through air-conditioning ducts, along water pipes and through walls. One of the most difficult buildings to search is one that has steam or hot water heat. This type of building will constantly thump, crack and tick due to the movement of the steam or hot water through the pipes and the expansion and contraction of the pipes. Additionally, the background noise of outside traffic, rain and wind can distort sound.

The person in charge of the room searching unit should look around the room and determine how the room is to be divided for searching and to what height the first searching sweep should extend. The first searching sweep will cover all items resting on the floor up to the selected height.

The room should be divided into two virtually equal parts. This equal division should be based on the number and type of objects in the room and not on its size. An imaginary line is then drawn between two objects in the room—e.g., the edge of the window on the north wall to the floor lamp on the south wall.

*First Room-Searching Sweep*

Look at the furniture or objects in the room and determine the average height of the majority of items resting on the floor. In an average room, this height usually includes table or desk tops and chair backs. The first searching height usually covers the items in the room up to the hip.

After the room has been divided and a searching height has been selected, both people go to one end of the room division line and start from a back-to-back position. This is the starting point, and the same point will be used on each successive searching sweep. Each person now starts searching around the room, working toward the other person, checking all items resting on the floor around the wall area of the room. When the two persons meet, they will have completed a "wall sweep." They should then work together and check all items in the middle of the room up to the selected hip height, including the floor under the rugs. This first searching sweep should also include those items that may be mounted on or in the walls, such as air-conditioning ducts, baseboard heaters and built-in wall cupboards, if these fixtures are below hip height.

The first searching sweep usually consumes the most time and effort. During all the searching sweeps, electronic or medical stethoscopes should be used on walls, furniture items and floors. (Figure 10-3).

*Second Room-Searching Sweep*

The person in charge looks at the furniture or objects in the room and determines the height of the second searching sweep. This height is usually from the hip to the chin or top of the head. The two persons return to the starting point and repeat the searching technique at the second selected searching height. This sweep usually covers pictures hanging on the walls, built-in bookcases and tall table lamps. (Figure 10-4).

*Third Room-Searching Sweep*

When the second searching sweep is completed, the person in charge again determines the next searching height, usually from the chin or the top of the head up to the ceiling. The third sweep is then made. This sweep usually covers high mounted air-conditioning ducts and hanging light fixtures. (Figure 10-5).

*Fourth Room-Searching Sweep*

If the room has a false, or suspended ceiling, the fourth sweep involves investigation of this area. Flush or ceiling-mounted light fixtures, air conditioning or ventilation systems, electrical wiring and structural frame members must be checked. A sign indicating SEARCH COMPLETED should be conspicuously posted in the area. If the use of a sign is not practical, a piece of colored Scotch tape should be placed across the doorjamb approximately 2 ft. above floor level. (Figure 10-6)

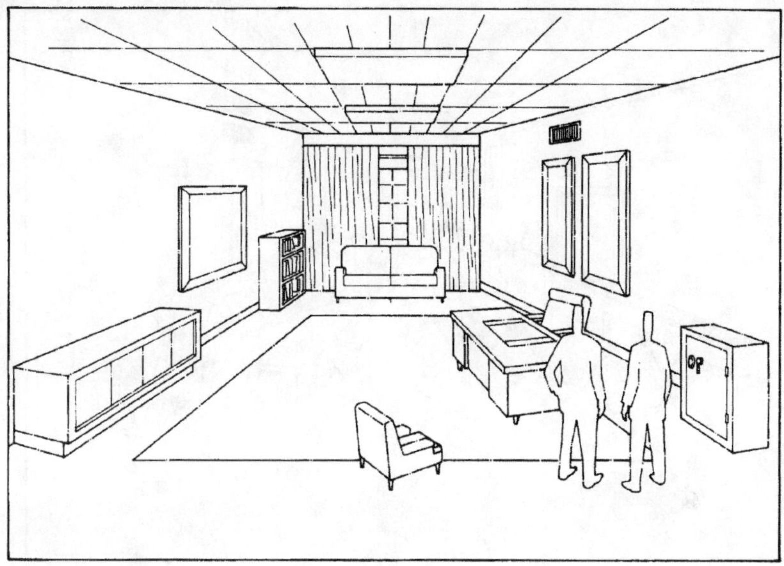

Figure 10-3. Step One. Room search—stop, listen. Courtesy: U.S. Department of the Treasury, Bureau of Alcohol, Tobacco and Firearms.

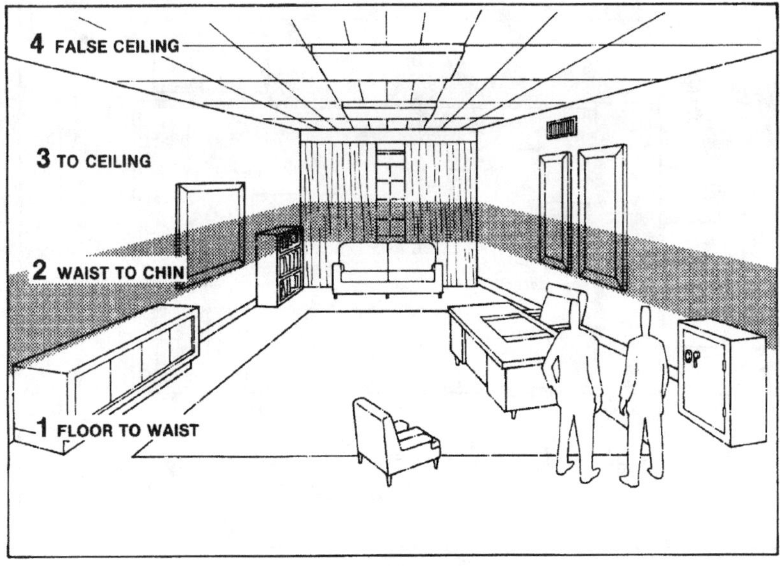


Figure 10-4. Step Two. Divide room by height for search. Courtesy: U.S. Department of the Treasury, Bureau of Alcohol, Tobacco and Firearms.

Figure 10-5. Step Three. Search room, by height and assigned area; overlap for better coverage. Courtesy: U.S. Department of the Treasury, Bureau of Alcohol, Tobacco and Firearms.

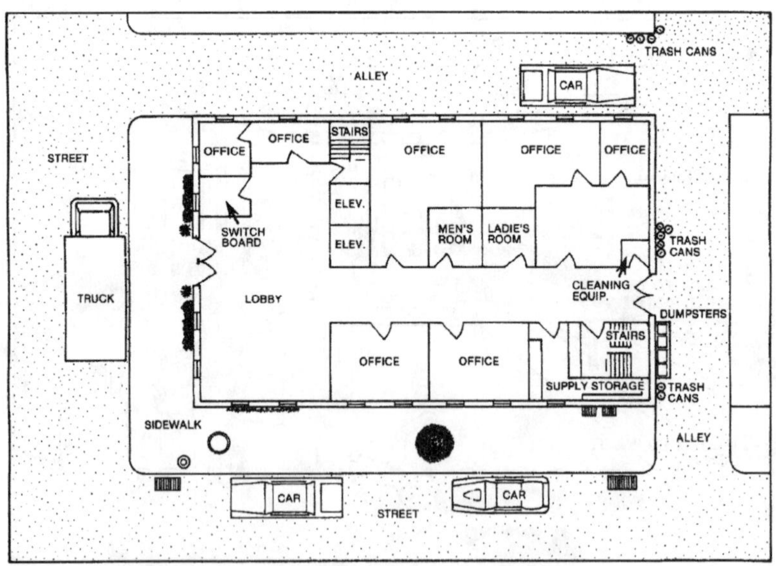


Figure 10-6. Step Four Search internal public areas. Courtesy: U.S. Department of the Treasury, Bureau of Alcohol, Tobacco and Firearms.

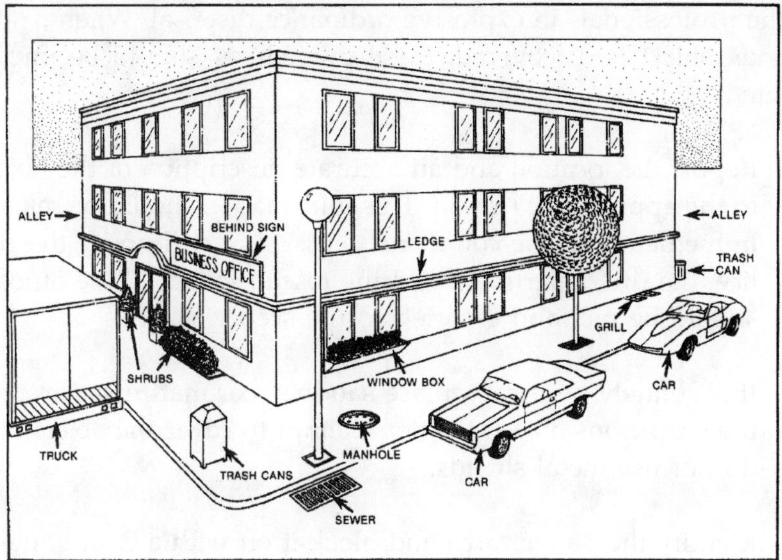



**Figure 10-7. Step Five. Search outside areas. Courtesy: U.S. Department of the Treasury, Bureau of Alcohol, Tobacco and Firearms.**

*Search Summary*

To summarize, the following steps should be taken in order:

1.   Divide the area and select a search height.
2.   Start from the bottom and work up.
3.   Start back-to-back and work toward each other.
4.   Go around the walls and proceed to the center of the room. The room searching technique can be expanded and can be used to search any enclosed area within the facility or building.

*Locating The Suspicious Object*

It is *absolutely imperative* that the personnel involved in a search be instructed that their only mission is to search for and report suspicious objects! Under no circumstances should anyone move, touch, or jar a suspicious object or anything that may be attached to it. The removal and disarming of a bomb must be left to the professionals in explosive ordinance disposal. When a suspicious object is discovered, these procedures should be implemented:

1.  Report the location and an accurate description of the object to the appropriate person. This information should be related immediately to the command center that will notify the police, the fire department and the rescue squad. These officers should be met and escorted to the scene.

2.  If absolutely necessary, place sandbags or mattresses around the suspicious object. Do not attempt to cover the object and do not use metal shields.

3.  Identify the danger area and block it off with a clear zone of at least 300 ft., including floors above and below the object.

4.  Check to see that all doors and windows are open to minimize primary damage from blast and secondary damage from fragmentation.

5.  Evacuate the building.

6.  Do not permit reentry into the building until the device has been removed and disarmed and the building has been declared safe for reentry.

## MEDIA RELATIONS

The issue of handling the media is of paramount importance in any bomb incident. All contacts with the media, including official responses, should be coordinated through a designated spokesperson. All other persons should be instructed not to discuss the situation with outsiders, especially the news media.

The purpose of this provision is to furnish the news media with accurate information and to see that additional bomb threat calls are not precipitated by irresponsible statements from uniformed sources.


## SUMMARY

Addressing the issue of bomb threats involves developing two separate but interdependent plans—the physical security plan and the bomb incident plan. Attention must be directed to those strategies that are designed to protect occupants and property. The physical security plan addresses the issue of prevention and control of access. The bomb incident plan provides the detailed procedures that must be employed when the bomb threat or actual bomb incident occurs.

### Sources

United States Department of The Treasury, Bureau of Alcohol, Tobacco And Firearms, *Arson and Explosives Incidents Report 1994*.

United States Department of The Treasury, Bureau of Alcohol, Tobacco And Firearms, *Bomb And Physical Security Planning*, ATF P 7550.2 (7-87).

United States Department of The Treasury, Bureau of Alcohol, Tobacco And Firearms, *Bomb Threat Checklist*, ATF F 7550.6 (10-93).

United States Department of The Treasury, Bureau of Alcohol, Tobacco And Firearms, *Suspect Package Alert*, ATF F 7550.6 (10-93).

Gustin, Joseph F. *Disaster And Recovery Planning: A Guide For Facility Managers*, 2nd Edition. Lilburn, GA.: The Fairmont Press, Inc., 2002.

# Chapter 11
# Evacuation

**W**hen a natural disaster occurs, a fire breaks out or somebody calls in a bomb threat, it may be necessary to evacuate the building. Orderly and complete evacuation of all occupants and visitors requires careful provision for egress routes and accounting for all individuals after the evacuation. With planning that takes into account all facts, quick and orderly evacuations can be achieved with minimal problems.

## EVACUATIONS

The nature of the specific disaster as well as the potential threat to occupant safety is the prime consideration that determines the order to evacuate.

A written evacuation plan is essential for two reasons. First, a written plan provides all building occupants with specific information on procedures for orderly evacuation in the event of a disaster or emergency. In addition, a written evacuation plan satisfies the mandate of the General Duty Clause of the OSH Act of 1970—to provide a safe and healthful working environment for all workers.

### Means of Egress

As defined in the *Life Safety Code Handbook* (6th edition, 1994) published by the National Fire Protection Association (NFPA), means of egress is the continuous and unobstructed way of travel from any point in a building or structure to a public way consisting of three separate and distinct components. These components are the:

- Exit access.
- Exit.
- Exit discharge.

### Exit Access

The exit access is that component of a means of egress that leads to an exit. For example, an exit access includes rooms and the building spaces that people occupy, as well as doors, corridors, aisle, unenclosed stairs and enclosed ramps that must be traveled to reach an exit.

### Exit

An exit is that component of the means of egress that is separated from all other spaces of a building by either constriction—which must have the minimum degree of fire resistance—or equipment so that a protected way of travel to the point of exit discharge is provided. Exits include exterior exit doors, exit passageways, separated exit stairs and separated exit ramps.

### Exit Discharge

The third component that comprises a means of egress is the exit discharge, or the path of travel from the termination of an exit to a public way. Since some building exits do not discharge directly into a public way, the path of travel may be within the building itself or outside the building. In either case, the purpose is to provide building occupants with the means for reaching safety.

Each of these components comprises the means of egress which must be provided from every location within the facility. Accordingly, all means of egress within the facility must be routinely inspected and regularly maintained to ensure maximum operability and utility by occupants seeking safety.

## Types of Evacuation

There are two types of evacuation—partial evacuation and total, or complete, evacuation. While the nature of the disaster/ emergency, as well as the potential threat to the safety of building occupants, determines the type of evacuation undertaken, the *immediacy* of the threat must also be considered.

### Partial Evacuation

In a partial evacuation, occupants who are immediately affected, or whose safety may be jeopardized by the occurrence, are relocated from the endangered area to a safe or secured area. This secured area may be either located within the facility itself or away from the building. For example, when fire is detected in a multistory facility or high-rise building, occupants on the floor immediately above and below the origin of the fire are evacuated.

### Total Evacuation

In a total or complete evacuation, all occupants are required to vacate the site premises with the possible exception of the disaster/emergency team members. In some situations, again depending upon the nature and immediacy of the disaster or emergency, team members may remain behind to ensure that all other occupants have vacated the site and/or to secure the building proper and critical building contents.

## Evacuation Factors

There are several factors that must be considered in determining which type of evacuation must be undertaken. These factors are:

- The nature of the disaster/emergency occurrence.

- The potential threat to occupant safety, as determined by the severity of the occurrence.

Each of these factors, in turn, determines the immediacy of the evacuation need.

*Evacuation: The Issue of Immediacy*

Facilities located in the Atlantic and Gulf coastal regions of the continental United States, as well as in the coastal areas of Hawaii and the Caribbean, are threatened by hurricanes. And while the losses incurred by hurricane damage can run into the hundreds of millions of dollars, the advance notice of hurricanes that is provided to the affected areas precludes, in many cases, personal injury

The National Weather Service's forecasting and tracking systems are able to identify and predict a hurricane's path and the storm's level of intensity with a fairly high degree of accuracy (Table 11-1). Based upon the available data received from these tracking systems, the National Weather Service issues notices to all areas within the storm's path, so that these areas have a fairly reasonable amount of preparation time. Often times, notice of an impending hurricane is provided so far in advance that a "formal" occupant evacuation becomes unnecessary. The "luxury" of this advance notice, therefore, provides facility managers and members of the disaster planning team with time to prepare for securing the building and its contents. Once an actual hurricane warning is issued by the National Weather Service, occupants should be instructed to leave the facility's premises by a given time, provided that such time is adequate for occupants to seek refuge on an individual basis. In situations such as these—where advance notice can be provided—the immediate need to evacuate is less urgent.

These notices usually take the form of "watches" and "warnings."

*Watch Versus Warning*

A watch is defined by those weather conditions that carry the potential for threat to safety. The warning, on the other hand, is defined in terms of actual conditions that exist and that threat is imminent. For example, a tornado warning may prompt an immediate partial evacuation of building occupants to pre-designated "safe" areas of refuge within the facility itself.

In the event of fire, either a partial or total evacuation may be ordered. While the floors immediately above and below the fire's location are generally evacuated first, there are occasions when all building occupants will be required to evacuate.

**Table 11-1. Hurricane scale. Source: National Weather Service.**

| Category | Wind Speed | Storm Surge |
|---|---|---|
| I | 74—96 mph | 4—5 ft. |
| II | 96—110 mph | 6—8 ft. |
| III | 111—130 mph | 9—12 ft. |
| IV | 115+ mph | 18+ft. |

*Other Evacuation Factors*

There are a number of other critical factors related to evacuation that must be addressed in developing evacuation strategies. These factors include determining:

- Internal authority/responsibility levels for offering and supervising the evacuation. Depending upon the nature and severity of the occurrence, the evacuation order may be made by the designated site person, the local fire department and/or other jurisdictional authority, or both.

- Timing of the order. Depending upon the potential threat to occupant safety, any decision to evacuate may need to be made immediately.

- Various types of assistance that will be made available to persons with disabilities. Assistance may come in the form of trained in-house personnel and/or fire and emergency services personnel, in assisting in the evacuation of occupants with disabilities.

Another critical factor that must be addressed in developing a facility's evacuation strategies is the threat of bombs and explosives-related violence.

## BOMB THREATS

The most serious of all decisions is whether to order an evacuation in the event of a bomb threat. According to the Bureau of Alcohol, Tobacco and Firearms, (ATF), there are three options available to a company's management team when a bomb threat is received. Facility managers may:

1.   Ignore the threat.
2.   Order an immediate and total evacuation.
3.   Search and evacuate if necessary.

### Option 1: Ignore the Bomb Threat

Ignoring a bomb threat completely can result in some problems. While a statistical argument can be made that few bomb threats are real, the fact that bombs have been located in connection with threats cannot be overlooked. If building occupants learn that bomb threats have been received and ignored, it could result in morale problems and have long-term adverse affect on business operations. In addition, there is the possibility that if the bomb threat callers feel that they are being ignored, they may actually go beyond the threat and, in fact, plant the bomb.

### Option 2: Order an Immediate Evacuation

Ordering an immediate and total building evacuation on every bomb threat received is, at face value, the preferred response. However, there are negative factors inherent in this option that must also be considered. For example, any immediate and total evacuation disrupts business operations. If bomb threat caller knows that a company's policy is to evacuate each time a bomb threat is received, they are likely to continue calling with the intention of bringing the facility's operations to a virtual standstill. Additionally, disgruntled or dissatisfied employees who know that their company's policy is to order an immediate and total evacuation may call in such threats in order to close the facility down. In doing so, the disgruntled employee is able to "exact revenge," or "set things straight." Another possible scenario is the student who wishes to cancel classes, avoid tests, etc., may call in a bomb threat. In addition, bombers who wish to cause personal injuries could place the explosive device near exits that are used for evacuation.

## Option 3: Search and Evacuate

Initiating a search after a bomb threat is received and then ordering the evacuation after finding a suspicious package or device is, perhaps, the most viable of all options, because it is not as disruptive as the immediate and total evacuation. It does satisfy the requirement to do something when the threat is received. Additionally, if an explosive device is discovered, an evacuation can be accomplished expeditiously while avoiding the potential danger areas of the bomb.

## EVACUATION AND SEARCH UNITS

In developing evacuation strategies for their own particular properties, facility managers and other members of the disaster response and recovery planning team should consider forming internal evacuation and search units. Comprised of selected management and supervisory personnel, this team's internal evacuation unit can direct the orderly evacuation and relocation of building occupants when such action becomes necessary. An evacuation unit that is well-trained in evacuation procedures becomes particularly important in the event of a bomb threat incident. Local police, fire and other jurisdictional authorities, as well as various community-based resources usually provide specific, bomb-incident evacuation training.

### Evacuation and Search Unit Training

In the event of a bomb threat, priority of evacuation becomes critical. The expeditious evacuation of occupants from the floor levels above and below the targeted or suspected area is necessary to remove occupants from danger.

There are specific site-search techniques that also comprise bomb incident training. This type of training, which is provided to members of the team's search unit, includes detection and location methods and procedures. Additional training that is provided to members of the team's search unit includes the proper procedures for marking and taping a room after that room or a particular area has been searched, and proper reporting procedures.

Once the bomb or incendiary device has been detected, it is imperative than any direct contact with the bomb be avoided. Its location should be well marked and a route back to the device should be noted. Indeed, one aspect of training that should not be included is techniques used to neutralize, remove or directly contact the explosive device. Only authorized enforcement officials with specialized knowledge in explosives should ever attempt to move, dismantle, or disengage a bomb.

For obvious reasons, only people who are thoroughly familiar with all aspects of a facility's floor plan should be recruited for the evacuation and search team. For a search to be effective, members of the search unit must have a thorough knowledge of hallways, restrooms, false ceilings and so on as well as any and all other site locations where explosive or incendiary devices may be planted or concealed. In many cases, local police and firefighters will be unfamiliar with the actual layout of the facility and will need to rely upon members of the search unit for building specifics.

## ADDITIONAL CONSIDERATIONS

There are several other considerations regarding evacuation that must be addressed in the disaster plan. These include the location to which occupants will report upon completion of the partial or total evacuation; confirmation of successful evacuation; and maintenance of evacuation routes.

### Reporting Locations

Regardless of the type of evacuation that may be ordered, a reporting location must be identified. In those situations that warrant a partial evacuation, consideration must be given to which area within the facility itself will be made available for evacuated occupants. Areas within the facility that have controlled or limited access for security reasons are not local choices. Neither are any business operations areas that may be disturbed by a sudden influx of additional personnel. The evacuation strategy must address location issues so that unnecessary problems do not crop up during an actual occurrence.

The evacuation strategy must also address the issue of total building evacuation. An off-site location must be identified. The location itself should provide safety from the disaster site and should be large enough to accommodate the total number of evacuees. Without an identified destination, the safety of evacuees could be jeopardized. Falling debris from the disaster site could seriously injure evacuees left standing or milling about the street. Additionally, the effectiveness of fire and emergency rescue services could be compromised by large numbers of people congregating outside the disaster site.

## Confirming Evacuation

Essentially, there are two methods by which an evacuation can be confined—the roll-call (or head count) method, and the search method. Each has its own distinct advantages depending upon occupancy type and facility size.

### The Roll-Call Method

For those facilities that house a relatively small number of regular occupants, the roll-call method is an effective means for verifying evacuation. The evacuation supervisor, or designee, can either count heads against the occupant list, or require a voice count against the occupant list to ensure that all occupants successfully evacuated.

### The Search Method

In larger facilities that house a relatively large number of occupants, the roll-call method is often not practical. In a total evacuation of these facilities wherein large numbers of regular occupants are either tenanted, or in those occupant types that include transients—i.e., mercantile, assembly occupancies—it becomes almost impossible to account for each person. In these instances, the search method is most effective.

Upon authorization from the appropriate jurisdictional agency to re-enter an evacuated site, the search method can be undertaken to determine successful evacuation. In those instances where authorization to re-enter an evacuated site is not immediately given, the jurisdictional agency may be called upon to conduct the post evacuation search.

## Maintaining the Evacuation Route

A final, but no less important consideration, is the issue of evacuation route maintenance. As part of a facility's routine inspection and maintenance program, special attention must be given to designated evacuation routes.

Essential from a regulatory standpoint, all routes designated and used in evacuation must be properly maintained. Additionally, all building components must be operable and ready for use in the event that an order to evacuate becomes necessary. These building components include lighting and emergency lighting systems, elevators, smoke and fire resistant ceiling and walls, and stair pressurization systems. And, finally, as previously discussed, all components that comprise the means of egress—the exit access, exit and exit discharge—must be properly maintained so that their utility can be ensure and that occupants can reach a point of safety.

## Sources

Cote, Ron, PE, Editor, *Life Safety Code Handbook,* 6th Edition, National Fire Protection Association, Quincy, MA, 1994.

*The Occupational Safety And Health Act* of 1970 (December 29, 1970), PL91-596,29 USC 651.

United State Department of The Treasury, Bureau of Alcohol, Tobacco And Firearms, *Bomb And Physical Security Planning,* ATF P 7550.2 (7087).

Gustin, Joseph F. *Disaster And Recovery Planning: A Guide For Facility Managers,* 2nd ed. Lilburn, GA: The Fairmont Press, Inc., 2002.

# Chapter 12
# *Fire/Life Safety*

## PREVENTION ISSUES AND RESPONSE MEASURES

While fire is a potential threat to any facility, there are measures that can be taken to minimize not only the threat of fire, but also the effects of fire should it occur. These measures should be included in any workplace fire safety program and incorporated into the overall disaster response and recovery plan. While these measures may be classified as preventive, they also serve as the basis for formulating a facility's specific response to fire emergencies.

These preventive and response measures involve assessing the building's general environment and those building components and features that are critical to effective fire protection. Preparing for the potential fire involves assessing a facility's fire readiness. Undertaking a comprehensive baseline survey of building systems and operations is the essential first step in determining fire readiness. The key points to consider include the:

- Fire protection systems and equipment, including control mechanisms.

- General operating work environment, including building systems and components.

## FIRE PROTECTION SYSTEMS

Fire protection systems include fire detection, alarms and communications systems. These systems, which are commonly referred to as protective signaling systems, serve three basic purposes. They are designed and intended to:

- Alert occupants to an emergency situation.

- Notify emergency personnel.

- Maximize "control" of facilities to minimize injury and protect property.

The type of fire detection and alarm system that is installed in a facility is contingent upon a number of factors, including:

- The building's occupancy classification (see Building Classifications later in this chapter).

- The mandatory code(s) established by the authority having jurisdiction.

For example, fire detection and alarm systems that are installed in facilities whose occupants are mobile and whose activities are either not restricted, or very casually monitored (e.g., in a business, retail, or educational occupancy), will have different requirements than detection and alarm systems that are installed in a facility whose occupants are restricted (e.g., hospital or correctional facility). The difference in required detection and alarm systems rests primarily with the degree of dependence upon others for survival. Facilities whose occupants are not restricted are minimally dependent upon other people for survival. Hospital patients who may be sedated, immobile, etc., as well as occupants of correctional facilities whose mobility is restricted are highly and, in many cases, totally dependent upon others for their survival.

However, while the type of required detection and alarm systems may vary, *all such systems* must provide for:

- Signaling (manual and/or automatic).

- Notification that action must be taken in response to a situation.

- Control, which activates the protection system themselves.

## Signaling

Signaling may include either manual fire alarm systems or automatic detection systems. In either case, signaling devices must be properly installed and maintained, as well as routinely inspected to ensure their operability and utility in the event of fire.

## Notifying Emergency Personnel

Detection and alarm systems must be equipped to notify both building occupants, fire brigades and other appropriate personnel including local fire departments, that an emergency does, indeed, exist. The facility's occupancy type may determine the appropriate means of notification.

For example, when an emergency situation occurs in a facility whose occupants are dependent upon others for their survival, only those people who are authorized to assist in relocation/ evacuation need to be notified.

While local codes, in conjunction with nationally recognized fire protection standards, determine the acceptability of visual and/or audio notification that is appropriate to occupancy type, another factor comes into play. The ADA determines, in large part, additional requirements for notification. The ADA's requirements, particularly Title III compliance provisions, mandate accommodation for people with disabilities. An occupant who is hearing-impaired, for example, must have accessibility to a notification system other than an audio notification system.

## Automatic Notification Systems

When a fire emergency occurs, the municipal fire department must be notified. The authority having jurisdiction, as well as other code requirements (NFPA 72, National Fire Alarm Code), determines the acceptability of automatic notification systems. According to the NFPA 72, National Fire Alarm Code, there are four acceptable methods of automatic notification delivery. They are:

1.   An auxiliary alarm system.
2.   A central station connection.

3.    A proprietary system.
4.    A remote station connection. □

*Auxiliary Alarm Systems*

Auxiliary alarm systems are those which are connected to the municipal fire alarm system. The alarm is transmitted on the same equipment and by the same methods as alarms that are manually transmitted from manual fire alarm boxes.

*Central Station Fire Alarm Systems*

Central station fire alarm systems are those which receive, record and maintain alarm signals at a central location, and whose personnel take appropriate action. Central station fire alarm systems are operated by companies who furnish, maintain and monitor supervised alarm systems.

*Proprietary Fire Alarm Systems*

Proprietary fire alarm systems are those which service contiguous and noncontiguous properties under one ownership and whose supervising station is located on-site. These systems are continually staffed by appropriately trained personnel.

*Remote Station Fire Alarm Systems*

Remote station fire alarm systems are those systems that transmit alarm signals from one or more protected premises to a remote location where appropriate action is taken.

**Control Mechanisms**

Control mechanisms are those which automatically engage the building system components that are necessary to protect occupants. These automatic control mechanisms include:

- Release of hold-open devices for doors and other openings.
- Automatic door unlocking.
- Stairwell and elevator shaft pressurization.
- Emergency lighting.
- Automatic sprinklers and standpipe systems.
- Other automatic extinguishing systems (e.g., dry chemicals,

low expansion foam, carbon dioxide and dry chemical extinguishing systems).

Because these various control mechanisms may not, in any way, impede or impair the operability of the fire detection and alarm components, continual and routine inspection and maintenance must be performed. Requirements for each control mechanism are determined by the authority having jurisdiction in accordance with NFPA code.

*Other Control Mechanisms*

Other control mechanisms include manual extinguishing equipment. Commonly known as portable fire extinguishers, this type of extinguishing mechanism is considered a universal in basic fire protection. Portable fire extinguishers are used in virtually all facilities and occupancy types. They serve as one of the first lines of defense in fire protection.

**Fire Extinguishers**

There are four general categories or classes of portable fire extinguishers:

- Class A Extinguishers, which are used on fires involving ordinary combustibles (e.g., paper, cloth or wood).

- Class B Extinguishers, which are used on fires involving flammable liquids (e.g., grease, gasoline and paints).

- Class C Extinguishers, which are used on fires involving electrical equipment.

- Class D Extinguishers, which are used on fires involving combustible materials.

Properly installed, inspected and maintained in accordance with NFPA standards, the effectiveness of portable fire extinguishers has been proved. However, there are several caveats regarding the use of portable fire extinguishers.

First, only those people who have been previously trained

in the proper use of extinguishers should attempt to use them in the case of actual fire. Untrained people can lose valuable time if they must stop to read the instructions. This lost time may not only minimize the utility of the portable fire extinguisher in containing the small fire, it may also render the extinguisher useless in many cases. Additionally, any time lost in containing a fire jeopardizes the safety of other building occupants, as well as the safety of the untrained person who attempts to use the portable extinguisher.

Second, if a fire cannot be extinguished quickly, containment efforts should be stopped. The person using the fire extinguisher should leave the fire location and seek refuge. Accordingly, if the fire is extinguished, the person operating the fire extinguisher should remain at the location until the fire department arrives.

Third, portable fire extinguishers should only be used when:

- All notification procedures have been activated.

- Occupant evacuation has begun.

- The fire is confined to a relatively small area.

- The trained person using the extinguisher has a clear escape route should the fire intensify or grow.

*Using The Portable Fire Extinguisher*

The procedure for using a fire extinguisher is basically the same, regardless of the extinguisher's class. Called the "PASS" procedure, it involves the following steps:

- Pulling the pin that unlocks the lever.

- Aiming the hose nozzle at the fire's base.

- Squeezing the lever to release the contents.

- Sweeping the flames from side to side, until they are extinguished.

**Assessing Fire Protection Systems**

As prevention issues, fire protection systems—including detection, alarm and communications systems—are critical to minimizing the effects of fire on a facility's occupants and on the property itself, as well as its contents.

To ensure the operability of a facility's fire protection system, the following questions should be addressed:

- Are all fire protection systems certified as required by local regulatory code as well as NFPA code?

- Are all fire protection systems, including alarms, tested annually at a minimum?

- If contained on premises, are all interior standpipes and valves tested regularly?

- If located on site property, are outside private fire hydrants flushed at least annually and on a routine preventive maintenance schedule?

- Are all fire doors and shutters in good operating condition?

- Are all fire doors and shutters unobstructed and protected against obstructions, including their counterweights?

- Are all fire door and shutter fusible links in place?

- Are automatic sprinkler system water control valves, air and water pressure checked weekly/periodically as required by code?

- Is the maintenance of automatic sprinkler systems assigned to appropriately trained people or to a sprinkler contractor?

- Are sprinkler heads protected by metal guards when exposed to physical damage?

- Is proper clearance maintained below sprinkler heads?

- Are portable fire extinguishers provided in adequate numbers and types?

- Are all portable fire extinguishers mounted in readily accessible locations?

- Are all portable fire extinguishers recharged regularly and appropriately noted on the inspection tags?

- Are employees periodically trained in the use of portable fire extinguishers and in all fire protection procedures?

- Is the local fire department familiar with the facility and its location and all specific hazards unique to the facility?

These questions serve as a basis for conducting the baseline survey discussed earlier. They are important for determining the effectiveness of the facility's fire protection system. And, in turn, it is a fire protection system's effectiveness that can ensure occupant safety and minimize damage and loss.

## GENERAL WORK ENVIRONMENT

Generally speaking a facility's purpose and function determine its occupancy. And occupancy patterns determine the general work environment. For example, the retail or mercantile facility's work environment is obviously different than the work environment of a hospital, factory, warehouse, or correctional facility. However, while the functions of facilities may vary, there are work environment issues, features, and building system components that are common to any facility regardless of its purpose or occupancy. Some of these common work environment issues include:

- Combustible scrap, debris and waste removal.

- Routine removal of accumulated combustible dust from elevated surfaces including the building's overhead structure.

- Prevention of interior combustible dust suspension through use of vacuum systems.

- Prevention of metallic conductive dust entering or accumulating on or around electrical enclosures and equipment.

- Use of covered metal containers for paint/chemically-soaked waste.

- Use of flame failure controls on oil/gas fired devices to prevent fuel flow if pilots and/or main burners become inoperable.

These issues, along with the features and building system components that are common to all facilities, are critical fire protection issues. As such, they should be addressed in terms of their effectiveness in the event of fire.

Some of the features and building systems that are common to all facilities include:

- Walkways.
- Floor and wall coverings.
- Stairs and stairways.
- Means of egress.
- Exit doors.
- Electrical system components.
- Flammable/combustible materials.

**Walkways**

Walkways are the travel paths that may be used in the event relocation and/or evacuation of occupants becomes necessary. Of critical importance during a fire, a walkway's accessibility should be assessed by determining if:

- Aisles and passageways are kept clear and unobstructed.

- Aisles and walkways are appropriately marked.

- Walking surfaces are covered with non-slip materials.

- Holes in floor coverings, walking surfaces are repaired properly.

- Changes in elevations or direction are readily identifiable.

- Adequate headroom is provided for the entire length of the aisle or walkway.

- Standard guardrails are provided where aisle or walkway surfaces are elevated more than 30 inches above any adjacent floor, or the ground.

## Floor And Wall Openings

Floor and wall openings are critical considerations in the event of fire. Proper maintenance of both kinds of openings must be ensured in order to not only prevent occupant injury during relocation and/or evacuation, but to prevent the spread of fire.

Floor and wall openings should be assessed by determining if:

- Required floor openings are guarded by a cover, guardrails, or equivalent protection on all sides, except at entrances to stairways or ladders.

- Grates or similar type covers over floor openings such as drains are designed and properly installed to ensure that foot traffic will not be affected by the grate spacing.

- Unused portions of service pits and pits not in use are covered or protected by guardrails or equivalent protection.

- Glass in windows, doors, glass walls, etc. that are subject to human impact are properly installed and maintained.

- Floor or wall openings in fire resistive construction are provided with doors or covers that are compatible with the fire rating of the structure and provided with self-closing features when appropriate.

## Stairs And Stairways

When an emergency situation requires the relocation and/or evacuation of occupants, stairs and stairways become one of the key focal points. Assessment of a facility's stairs and stairways includes determining if:

- Standard stair rails or handrails are installed on all stairways having four or more risers.

- All stairways used for purposes of fire escape in existing buildings are at least 22 inches wide.

- Stairs have landing platforms not less than 30 inches in width at every 12 ft. or less of vertical rise.

- Stair angles are not more than 50 degrees and not less than 30 degrees.

- Stairs of hollow-pan type treads and landings are filled to the top edge of the pan with solid material.

- Step risers on stairs are uniform from top to bottom.

- Steps on stairs and stairways are designed or provided with a surface that renders them slip resistant.

- Stairways handrails are located between 30 and 34 inches above the leading edge of stair treads.

- Stairway handrails have at least 3 inches of clearance between the handrails and the surface on which they are mounted.

- Doors or gates that open directly on a stairway are provided with platforms so that the swing of the door does not reduce the width of the platform to less than 21 inches.

- Stairway handrails are capable of withstanding a load of 200 pounds, applied within 2 inches of the top edge, in any downward or outward direction.

- Stairs or stairways that exit directly into areas of traffic are equipped with adequate warnings designed to prevent occupants from stepping into the path of traffic.

- Stairway landings have a dimension measured in the direction of travel at least equal to the width of the stairway.

- The vertical distance between stairway landings is limited to 12 ft. or less.

**Means of Egress**

In the event of a fire emergency, a facility's means of egress, or exit, enable occupants to evacuate their immediate work area and move towards safe refuge. Assessing the effectiveness of means of egress includes determining if:

- All exits are marked with an exit sign and illuminated by a reliable light source.

- The directions to exits, when not immediately apparent, are marked with visible signs.

- Doors, passageways or stairways that do not serve as either exits or access to exits but which could be mistaken for exits, are appropriately marked NOT AN EXIT, TO BASEMENT, STOREROOM, etc.,

- Exit signs are provided with the word EXIT in lettering at least 6 inches high and the stroke of the lettering at least 3/4-inch wide.

- All exit doors are side-hinged.

- All exits are kept free of obstructions.

- Two means of egress, minimally, are provided from elevated platforms, pits or rooms where the absence of a second exit would increase risk of injury from hot, toxic, corrosive, suffocating, flammable, or explosive substances.

- There are sufficient exits to permit prompt escape.

- The number of exits from each floor of a building and the number of exits from the building, itself, is appropriate for the building occupancy load.

- Exit stairways which are required to be separated from other parts of a building are enclosed by at least 2-hour fire-resistive construction in buildings more than four stories in height and not less than 1-hour fire-resistive construction elsewhere.

- The slope of all ramps used as part of required exiting from a building is limited to 1 ft. vertical and 12 ft. horizontal.

- Any frameless glass doors, glass exit doors, etc. used for exiting are fully tempered and meet all applicable safety requirements and standards for human impact.

**Exit Doors**

The issue of exit doors is critical in a fire emergency. As an integral component in the relocation and/or evacuation effort, exit doors serve as a primary means for occupant escape. Therefore, the issue of exit doors in a facility should be assessed by determining if:

- The design, construction and installation indicate obvious and direct exit travel.

- Windows which could be mistaken for exit doors are made inaccessible by means of barriers or railings.

- Exit doors can be opened from the direction of exit travel without the use of a key or any special knowledge or effort when the building is occupied.

- Any revolving, sliding overhead door is prohibited from serving as a required exit door.

- Panic hardware installed on required exit doors will allow opening when a force of 15 pounds or less is applied in the direction of exit travel.

- Adequate barriers and warnings are provided for those exit doors that open directly onto any street, alley, or other area where vehicles are operated.

## ELECTRICAL SYSTEM COMPONENTS

The lifeline of any building is its electrical system. Electricity not only drives building and business components and operations, but electrical systems also control them. Electrical energy carries with it the potential for danger, including the danger of fire.

To assess not only the effectiveness of a facility's electrical system, but also its danger potential, all system components must be routinely inspected and maintained. Evaluating the electrical system's effectiveness, as well as its danger potential, includes determining if:

- Occupants are required to report any obvious hazards in connection with electrical systems and/or components.

- All electrical equipment and tools are either grounded or double-insulated.

- All electrical appliances, including vacuuming systems, polishers, etc., are grounded.

- All extension cords in use have grounding conductors.

- Multiple plug adapters are prohibited, and if so, are they still being used.

- Ground-fault circuit interrupters are installed on each temporary 15 or 20 ampere, 120V AC circuit at locations where construction, demolition, modifications, alterations, or excavations are being performed.

- All temporary circuits are protected by suitable disconnection switches or plug connectors at the junction with permanent wiring.

- Electrical installations present in hazardous dust and vapor areas meet the National Electrical Code (NEC) for hazardous locations.

- All cord, cable and raceway connections are intact and secure.

- All flexible cords and cables are free of splices and/or taps.

- All interior wiring systems include provisions for grounding metal parts of electrical raceways, equipment and enclosures.

- All disconnecting switches and circuit breakers are labeled to indicate their use, or all unused openings in electrical enclosures and fittings are closed with appropriate covers, plugs, or plates.

- All electrical enclosures (e.g., switches, receptacles, junction boxes, etc.) are provided with tight-fitting covers or plates.

- Disconnecting switches for electrical motors in excess of two horsepower are capable of opening the circuit without exploding, when the motor is stalled.

**Flammable and Combustible Materials**

Virtually ever facility has materials on site than can be classified as either flammable and/or combustible. Ranging from the generic chemical solvents that are used for cleaning, to materials that are used in various office operations (e.g., dry chemicals used for reproduction), and to industry and/or facility specific materials (e.g., chemicals, petro-chemicals, agri-chemicals, etc.), facility managers must be aware of the potential threat to safety that these stored energy sources present.

In order to assess the various levels of protection, as well as the danger potential that these stored contents present, it is necessary to determine if:

- All combustible scrap, debris and waste materials are stored in covered metal receptacles and promptly and properly removed.

- Proper storage is utilized to minimize the risk of fire including spontaneous combustion.

- Only approved containers and tanks are used for the storage and handling of any and all flammable and combustible liquids.

- All connections on drums and combustible liquid pilings, vapors and liquids are tight.

- All flammable liquids are kept in closed containers when not in use.

- Bulk drums of flammable liquids are grounded and bonded to containers during dispensing.

- All storage rooms for flammable and combustible liquids have explosion-proof light fixtures.

- All storage rooms for flammable and combustible liquids have mechanical or gravity ventilation.

- Liquefied petroleum gas that is stored, handled and used is done so in accordance with industry practice, code and standards.

- NO SMOKING signs are posted on liquefied petroleum gas tanks.

- All solvent wastes and flammable liquids are kept in fire-resistant, covered containers until they are removed from the site.

- Vacuuming is used, rather than blowing or sweeping of combustible dust whenever possible.

- Firm separators are placed between containers of combustibles or flammables, when stacked, to assure support and stability.

- Fuel gas cylinders and oxygen cylinders are separated by distance, fire resistant barriers, etc., while in storage.

- Fire extinguishers are selected and provided for the types of materials in areas where they are to be used—i.e., Class A—ordinary combustible material fires; Class B—flammable liquid, gas or grease fires; Class C—energized electrical equipment fires; Class D—combustible metal fires.

- Appropriate fire extinguishers are mounted within 75 feet of outside areas containing flammable liquids, and within 10 feet of any inside storage area for such materials.

- Extinguishers are free from obstructions.

- All extinguishers are serviced, maintained, and tagged at intervals not to exceed one year.

- All extinguishers are fully charged and in their designated places.

- The nozzle heads of permanently mounted sprinkler systems are so directed or arranged that water will not be sprayed into operating electrical switchboards and equipment.

- NO SMOKING signs are posted where appropriate in areas where flammable or combustible materials are used or stored.

- Safety cans used for dispensing flammable or combustible liquids are at a point of use.

- Maintenance procedures call for the immediate clean up of flammable or combustible liquid spills.

- Storage tanks are adequately vented to prevent the development of excessive vacuum or pressure as a result of filling, emptying, or atmosphere temperature changes.

- Storage tanks are equipped with emergency venting that will relieve excessive internal pressure caused by fire exposure.

- NO SMOKING rules are enforced in areas involving storage and use of hazardous materials.

As part of a facility's general work environment, each of these features and building system components becomes critical fire protection issues. As such, they should be assessed in terms of their effectiveness in the event of fire.


## BUILDING CLASSIFICATIONS

A facility's classification is determined in large part by jurisdictional code in conjunction with standard-setting organizations. Depending upon its purpose, function and use, as well as its occupancy type, a building can fall into various categories. These categories, or classifications, as defined by the National Fire Protection Association:

- Assemblies
- Correctional and detention facilities
- Education
- Business
- Healthcare
- Mercantile
- Residential
- Industrial
- Storage

### Assemblies
An assembly is a building/facility, or a portion of a building/ facility, whose purpose is to accommodate 50 or more people at a given time. Examples of assembly occupancies include:

- Armories

- Assembly halls

- Auditoriums (including exhibition halls and gymnasiums)

- Club rooms

- Conference rooms and college/university classrooms that have an occupant capacity of 50 or more people

- Courtrooms

- Dance halls

- Restaurants, drinking establishments and nightclubs (with a 50+ occupant load)

- Recreation centers

- Libraries and museums

- Theaters

- Places of worship

- Public transportation facilities and terminals, including air, surface, sub-surface and marine facilities

**Business**

Places of business are those which are used for conducting the normal activities of a business, including general operations such as finance, administration, etc. Examples of places of business include:

- Colleges and universities' instructional buildings and classrooms with an occupant load of less than 50 people, and instructional laboratories

- Professional services including doctors and dentists' offices and outpatient clinic

- General offices

- Municipal facilities, including town halls, courthouses, etc.

**Correctional and Detention**

Correctional and detention facilities are those whose occupants are incarcerated and/or detained. Because the degree of control over occupant activity is extreme, occupants in almost all cases are completely dependent upon others for their safety from fire. Examples of correctional/detention facilities include:

- Adult and juvenile detention facilities
- Adult and juvenile correctional facilities
- Adult and juvenile work camps
- Juvenile training schools

## Education

Educational facilities include buildings portions of buildings that are used for educational purposes through the 12th grade, by six or more individuals for four or more hours per day, or for more than 12 hours per week. Educational facilities include:

- Academies
- Kindergartens
- Nursery schools
- Elementary and middle schools, junior and senior high schools

## Healthcare

Healthcare facilities are those which are used for providing medical and related services to occupants whose illness, disability or infirmity may make them incapable of protecting themselves in a fire emergency. Examples of healthcare facilities include:

- Hospitals and nursing homes
- Limited use facilities
- Ambulatory healthcare centers

## Industrial

Industrial facilities include those that are used for the manufacturing of various products as well as those whose operations include assembling, decorating, finishing, packaging, processing, repairing, etc. Occupants may be exposed to a range of hazardous materials and processes. Examples of industrial facilities include:

- Dry cleaning plants
- General factories
- Food processing plants
- Laundries
- Power plants
- Pumping stations
- Refineries

## Residential

Residential facilities are those which provide sleeping accommodations and include all buildings designed for such purposes. Examples of residential facilities include:

- Hotels, motel and dormitories
- Apartment buildings
- Lodging/rooming houses
- Boarding and care facilities

## Mercantile

Mercantile facilities are those which include stores, markets, etc., and those that are used for the display and sale of merchandise. Examples of mercantile facilities include:

- Auction rooms
- Department stores
- Drug stores
- Shopping centers
- Supermarkets

## Storage

Storage facilities are those buildings used for storing goods, merchandise, vehicles or animals. Examples of storage facilities include:

- Barns
- Bulk oil storage
- Cold storage

- Freight terminals
- Grain elevators
- Parking structures
- Stables
- Truck and marine terminals
- Warehouses.

Since there are different code requirements for each classification, building owners and facility managers should exercise caution in determining the appropriate occupancy classification for their particular site(s). Consulting with the appropriate local jurisdictional authority can prevent any improper classification of building and occupancy type, thus ensuring that fire safety measures are not compromised.

A comprehensive and through fire safety plan must also consider other factors. Among the most significant of these factors is hazard classification.

## HAZARD CLASSIFICATION

In its *Life Safety Code Handbook*, the NFPA addresses the issue of hazard contents. While the Association's hazard classification is based on the potential threat to life, as that potential is presented by a building's contents, its method of classification is based on life safety. Thus, classification of hazard contents is based on the threat of fire, explosion and other similar occurrences.

The NFPA's classification of hazard contents follows:

**Low Hazard**—Contents which are of such low combustibility that self-propagating fires cannot occur.

**Ordinary Hazard**—Contents are those which are likely to either burn with moderate rapidity or which emit considerable smoke.

**High Hazard**—Contents are those which are most likely to burn with either extreme rapidity or which are most likely to cause explosions.

Despite the fact that the contents of most occupancies are

classified as ordinary hazards, there are certain procedures that all building occupants must follow, if and when a fire or other emergency occurs. These procedures which are essential in ensuring the safety of all building occupants and which should be clearly outlines in the occupants' emergency procedures manual include:

- Notifying the facility manager or other appropriate internal contact.
- Notifying the local fire department.
- Activating the manual alarm system.

In addition, specific procedures that outline the steps required for emergency evacuation and/or relocation must be clearly spelled out in the *Occupants' Emergency Manual*. Essential to ensuring the safety of occupants, procedures should include the following "universal" steps:

1.  Occupants should be instructed to close but not lock doors behind them as the leave.

2.  Doors should be touched prior to being opened. A hot door indicates fire on the opposite side, and the door should not be opened.

3.  Stairway doors should be kept closed except when people are moving through them. Holding doors open will cause smoke to be drawn into the stairwell.

4.  If smoke is encountered, occupants should breathe through a handkerchief or piece of clothing to reduce smoke inhalation.

5.  If caught in heavy smoke, people should drop to their hands and knees and crawl. People should hold their breath as much as possible.

6.  If clothing catches fire, people should stop, drop and roll. Attempting to run will only fan the flames and spread the fire.

7. If people become trapped in a room, the doors should be closed and the door sill should be covered with a towel or other object to limit smoke infiltration. People should attempt to move to a perimeter area and signal for help from a window.

8. Windows should not be broken out except as a last resort. Breaking a window may cause smoke infiltration from within the building due to pressure differentials or from smoke rising up the side of the building.

**Sources**

Cote, Ron, PE, Editor *Life Safety Code Handbook*, 6th Edition, National Fire Protection Association, Quincy MA, 1994.

Gustin, Joseph F. *Disaster & Recovery Planning: A Guide for Facility Managers*, 2nd Edition, Lilburn, GA: The Fairmont Press, Inc., 2002.

Gustin, Joseph F. *Safety Management: A Guide For Facility Managers*, New York: UpWord Publishing, Inc., 1996.

# Appendix I
# *ADA Checklist for New Lodging Facilities*

## I. PARKING AND LOADING ZONE

A. **I**f "self parking" is provided, are at least the minimum number of accessible parking spaces provided as required in surface lots or parking garages, including "van accessible" spaces for those who use lift-equipped vans? (ADA Standards 4.1.2(5))

B. Does each accessible parking space have or share an adjacent access aisle to allow persons who use wheelchairs, walkers or other mobility aids to transfer from their car/van? (ADA Stds. 4.6.3)

C. Do the parking spaces and access aisles have surface slopes less than 1:50 (i.e.: is the critical dimension 1/2 inch or less? (ADA Stds. 4.6.3)

D. Does each accessible parking space have a post- or wall-mounted sign with the symbol of accessibility mounted high enough so the sign is visible when a vehicle is parked in the space? (ADA Stds. 4.6.4)

E. Are the level surfaces of accessible parking spaces and access aisles free of "built-up" curb ramps so persons who use mobility aids (e.g.: wheelchairs, walkers or crutches) can make convenient transfers? (ADA Stds. 4.6.3)

F.   If there is more than one accessible parking space, are the accessible parking spaces the closest parking spaces to the lobby entrance and accessible guestroom entrance(s)? (ADA Stds. 4.6.2)

G.   If the lodging facility has covered passenger pickup/drop-off areas, does the pavement at such area(s), including the required 5' wide access aisle, slope 1:50 or less (critical dimension of 1/2 inch or less)? (ADA Stds. 4.6.6)

H.   Is the height of the covered passenger pickup/drop-off area at least 9'-6" to allow vans with raised roofs to use such area? (ADA Stds. 4.6.5)

I.   Does the parking garage serving the lodging facility allow at least 98" of vertical clearance for vehicles with raised roofs to approach, use and exit the accessible parking spaces? (ADA Stds. 4.6.5)


## II.   EXTERIOR ROUTES

A.   Regarding the exterior routes (e.g.: sidewalks and walkways). Are there no steps, no abrupt level changes over 1/4 inch, and no unramped curbs that will impede access for persons who use wheelchairs, walkers and other mobility aids between...
   1.   ...the accessible parking space access aisles (both guest and employee parking areas) and an accessible entrance door to each building? (ADA Stds. 4.1.2(1))
   2.   ...the accessible parking space access aisles and the exterior doors to the accessible guestroom(s)? (ADA Stds.4.1.2(1))
   3.   ...the accessible passenger pick-up/drop off area and an accessible entrance door? (ADA Stds.4.1.2(1))
   4.   ...the lobby and the accessible guestrooms with exterior room doors? (ADA Stds. 4.1.3(1))
   5.   ...the accessible guestroom(s) and exterior amenities,

such as swimming pools, whirlpools, dressing areas, restrooms, picnic areas, bars and outdoor dining areas? (ADA Stds. 4.1.2(2))

6. ...the public sidewalk (if provided) or street, and an accessible entrance door to the lobby? (ADA Stds. 4.1.2(1))

7. ...an accessible entrance to the facility and the public transportation stops serving this site? (ADA Stds. 4.1.2(1))

8. ...ground level fire exit doors (including any at loading docks) and a driveway, public sidewalk, street or other "public way"? (ADA Stds. 4.1.3(9))

B. Examine each of the exterior routes described above, to confirm the following:

1. ...do the sidewalks have cross slopes (i.e.: slopes tilting side to side) less than 1:50 (critical dimension of 1/2 inch or less) to persons who use wheelchairs can conveniently negotiate the routes? (ADA Stds. 4.3.7)

2. ...excluding ramps described below, do the sidewalks have running slopes (i.e.: in the direction of travel) that are no greater than 1:20 (1-1/4 inch critical dimension or less)? (ADA Stds. 4.3.7)

3. ...is the usable width of sidewalks at least 36" wide to accommodate wheelchair travel, even if cars project over the curb onto the sidewalk? (ADA Stds. 4.3.3)

4. ...are the curb ramps at least 36" wide, excluding the flared sides, to allow for convenient wheelchair travel? (ADA Stds. 4.7.3)

5. ...do the curb ramps have running slopes of 1:12 or less (critical dimension of 2" or less)? (ADA Stds. 4.7.2)

6. ...are the other exterior ramps at least 35" wide (between handrails)? (ADA Stds. 4.8.3.)

7. ...do the exterior ramps have running slopes of 1:12 or less (critical dimension of 2" or less)? (ADA Stds. 4.8.2)

8. ...do the gratings on the sidewalks and walkways have spaces no more the 1/2" wide in the direction of travel

so that canes, crutches and walkers do not skip into them, causing an individual to fall? (ADA Stds. 4.5.4)

9.   ...do the exterior ramps have the following features:

    a.   top and bottom landings that are level, at least as wide as the ramp they serve, and at least 60" long to allow for adequate maneuvering and resting space for persons who use wheelchairs, walkers, and other mobility aids? (ADA Stds. 4.8.4)

    b.   if a ramp is more than 30' long, is there a middle landing that is level, at least as wide as the ramp run it serves and at least 60" long? (ADA Stds. 4.8.4)

    c.   are the exterior ramps at least 36" wide between the two handrails to allow for convenient wheelchair travel? (ADA Stds. 4.8.3)

    d.   do the exterior ramps have running slopes that are 1:12 or less (critical dimension of 2" or less)? (ADA Stds. 4.8.2)

10.  ...are all exterior stairs built so blind persons and persons with low vision will not hit their heads on the underside (i.e.: protected with a cane detectable warning, such as a planter or enclosed with walls so a continuous 80" high circulation path is provided for building users)? (ADA Stds. 4.4.2)

11.  ...since blind persons and persons with low vision can walk on any sidewalks, are all sidewalks and walkways free of any objects (i.e.: fire extinguishers, wall mounted lights, electrical meters, signs, pay phones, trees, shrubs, etc.) that pose a hazard to blind persons and persons with low vision by projecting into the path more than 4"? (ADA Stds. 4.4; 4.1.2(3); 4.1.3(2))


### III.  BUILDING ENTRANCES & LOBBY

A.   If a fully automatic door is not provided, is the walkway in front of the lobby door level, without any portion steeper than 1:50 (critical dimension of 1/2" or less), so persons who use wheelchairs do not roll away from the door when they take their hand off the wheelchair and reach for the door hardware? (ADA Stds. 4.13.6)

B.  Does at least one lobby entrance door allow at least 32" clear passage width so persons who use wheelchairs, walkers, crutches, and other mobility aids can get through the door? (ADA Stds. 4.13.5)

C.  Is the door hardware (lever, pull, panic bar, etc.) usable with one hand, without tight grasping, pinching or twisting of the wrist, since many persons with disabilities may not have high manual dexterity or use of both hands? (ADA Stds. 4.13.9)

D.  If there a vestibule without fully automatic doors, is there a 30"x48" clear floor space where one can be outside the swing of a hinged door (i.e.: for out-swinging doors, at least 7'-0" between the exterior door frame and interior door frame) to allow persons who use wheelchairs to proceed through one door without it closing on them and binding the wheelchair as they approach and open the next door? (ADA Stds. 4.13.7)

E.  If there is key card controlled door hardware on building entrances, is the key card reader positioned so persons who use wheelchairs may approach and operate the opener (48" high maximum if only front approach, 54" high if parallel approach is available)? (ADA Stds. 4.1.3(10); 4.27.3; 4.2.5; 4.2.6)

F.  Do the registration counters or other counters serving guests have a lowered portion no more than 36" high or is there a folding shelf at 36" high to allow persons who use wheelchairs to fill out registration forms? (ADA Stds. 7.2)

G.  If a counter is used for serving breakfast or other food products, does it have at least a 36" long section that is no higher than 36" above the floor to allow persons who use wheelchairs to reach self-serve items? (ADA Stds. 7.2)


## IV.  INTERIOR ROUTES

A.  **Abrupt Level Changes**—Are hallways and corridors free of any steps or abrupt vertical level changes over 1/4" that will

impede access for persons who use wheelchairs, walkers, and other mobility aids between... (ADA Stds. 4.1.3.(1))
...the lobby and accessible guestrooms, suites, and other sleeping rooms?
...the lobby and any restaurants, other dining areas, and vending/ice machine areas?
...the lobby and the public and employee restrooms?
...the lobby and gift shops, newsstands, or other retail shops in the facility?
...the lobby and all other guest amenities such as, exercise/recreational areas, interior pools, business centers, and guest laundry rooms?
...the accessible entrances and all employee only work areas (e.g.: back of registration counter, housekeeping storage rooms, kitchens, administrative offices, etc.)

B.   **Slopes**—Examine each of the interior routes described above:
...excluding accessible ramps, do the corridors have running slopes (i.e.: in the direction of travel) not more than 1:20 (critical dimension of 1-1/4" or less)? (ADA Stds. 4.3.7; 4.8)
...are the interior ramps at least 36" wide between the two handrails to allow for convenient wheelchair travel? (ADA Stds. 4.8.3)
...do the interior ramps have running slopes that are 1:12 or less (critical dimension of 2" or less)? (ADA Stds. 4.8.2)
...do the interior ramps have top, middle and bottom landings that are level and at least 60" long to allow for adequate maneuvering and resting space for persons who use wheelchairs, walkers, and other mobility aids? (ADA Stds. 4.8.4)
...if a ramp is more than 30' long, is there a middle landing that is level, at least as wide as the ramp run it serves, and at least 60" long? (ADA Stds. 4.8.4)
...for corridors into which at least one door swings, is there at least 54" clear floor space opposite the door, to allow a person who uses a wheelchair adequate space to turn and enter the doorway? (ADA Stds. 4.13.6)

C.  **Doors**—Examine each of the interior routes described above:
    ...with the exception of doors at shallow closets, do the doors to required accessible spaces, and all doors into and within guestrooms allow at least 32" clear passage width for wheelchairs, crutch users, and persons who use walkers (ADA Stds. 4.1.3.(7); 4.13.5;9.4)
    ...with the exception of non-accessible guestrooms, does at least one door to each accessible space have door hardware (levers, pulls, panic bars, etc.) that is usable with one hand, without tight grasping, pinching, or twisting of the wrist, since many persons with disabilities may not have high manual dexterity or use of both hands? (ADA Stds. 4.13.9)
    ...with the exception of guestrooms not designated as accessible, does at least one door to each accessible space have at least 18" of clear floor space on the latch side for persons who use wheelchairs, walkers and other mobility aids to approach and pull open? (ADA Stds. 4.13.6)

D.  **Protruding Objects**—Examine all interior hallways, stairways and other pedestrian routes:
    ...to minimize the risks to blind persons and persons with low vision, are all of these areas free of objects (i.e.: fire extinguisher, wall mounted lights, electrical meter, sign, pay phone, etc.) mounted between 27"—80" high, that project into the path more than 4"? (ADA Stds. 4.4)
    ...are the interior stairs built so blind persons or persons with low vision cannot hit their heads on the underside? (ADA Stds. 4.4.2)

E.  **Elevators**—If the facility has more than 2 stories, including any basement levels, is there a full size passenger elevator serving each level of the hotel, including the basement for persons with disabilities who cannot use stairs? (ADA Stds. 4.1.3(5))
    Examine passenger elevators for the following...
    ...are all of the elevator lobbies free of ash trays or other elements placed below the elevator call buttons that project

more than 4"? (ADA Stds. 4.10.3.)

...are all elevator jambs provided with signs placed on both sides designating the floor with 2" minimum height raised letters and Braille characters centered at 60" above the finish floor? (ADA Stds. 4.10.5)

...are all of the elevators equipped with audible tones/bells or verbal annunciators that designate the passage of floors? (ADA Stds. 4.10.13)

...are all of the elevators equipped with audible tones/bells or verbal annunciators that designate the direction of the elevator called—one tone for "up" and two tones for "down"? (ADA Stds. 4.10.4)

...are all the elevators with an emergency communication system equipped with a system that does not require only voice communication (i.e.: either TTY systems or a system of lights and signs designating the meaning of the lights)? (ADA Stds. 4.10.14)

...if the elevators have emergency communication systems behind cabinet doors, is the door pull usable with one hand, without tight grasping, pinching or twisting of the wrist, since many persons with disabilities may not have high manual dexterity or use of both hands? (ADA Stds. 4.10.14)

...are the highest floor control buttons in the elevator(s) mounted within 54" of the floor, with associated raised letters and Braille characters? (ADA Stds. 4.10.12)

F.   **Stairs**—Are there floors connected only by stairs, not an elevator or ramp? (ADA Stds. 4.1.3.(4))

If yes, then...

...do the stairs have closed risers so a crutch or cane cannot skip through the open space between the stair risers and people will not lose their balance as they climb the stairs? (ADA Stds. 4.9)

...are all stair treads the same depth (at least 11" deep), measured riser to riser, to prevent tripping hazards for persons with disabilities? (ADA Stds. 4.9)

...are there handrails on both sides of the stair at a uniform height of 34"-38" above the front edge of the step? (ADA Stds. 4.9)

...do the handrails extend horizontally at the top and bottom of each stair section to give persons who have difficulty using the stairs a stable gripping location before ascending or descending? (ADA Stds. 4.9)

...are the handrails continuously grippable, without interruption by vertical supports, newel posts, or other construction elements which require repositioning the hands while ascending or descending? (ADA Stds. 4.9)

G.   **Areas of Rescue Assistance**—In hotels that do not have a supervised fire sprinkler system serving every room, are there two 30"x48" wheelchair waiting areas (a.k.a. and "area of rescue assistance" linked to the primary entry by intercom) at each required exit (i.e.: stairs) on levels above or below the ground floor for persons with disabilities who cannot exit the building in case of a fire or emergency? (ADA Stds. 4.1.3(9))

H.   **Drinking Fountains**—Are at least 50% of the drinking fountains on each floor mounted so the spout is no higher than 36" (ADA Stds. 4.15.3.)

I.   **Public Telephones**—Are at least the following accessible telephone elements provided per facility... (ADA Stds. 4.1.3(17))

1.   If one public pay phone or bank of pay phones is provided on a given floor, does at least one pay phone have the flowing features:

   a.   Is it mounted with the coin slot no higher than 54" above the floor?

   b.   Does the accessible phone have volume controls?

   c.   Do at least 25% (not less than one) of all other pay phones on each floor have volume controls?

   d.   If the bank of phones includes at least 3 pay phones, is there a shelf and an electrical outlet to allow for TTY (text telephone) use by persons who are deaf?

    2.    If more than one bank of public pay phones is provided on a given floor, does at least one pay phone per bank have the following features?

        a.    Is it mounted with the coin slot no higher than 54" above the floor, and one pay phone on that floor with the highest operable element no higher than 48" above the floor?

        b.    Do the accessible phones have volume controls?

        c.    Do at least 25% (not less than one) of all other pay phones on each floor have volume controls?

        d.    If one or more banks of phones include at least 3 pay phones, is there a shelf and an electrical outlet to allow for TTY (telephone typewriter) use by persons who are deaf?

    3.    If one house phone or one bank of house phones is provided on a given floor, does at least one house phone have the following features: (ADA Stds. 4.1.3.(17))

        a.    Is it mounted with the handset cradle no higher than 54" above the floor?

        b.    Does the accessible phone have volume controls?

        c.    Do at least 25% (not less than one) of all other house phones on each floor have volume controls?

    4.    Is there a sign at each single pay phone or pay phone bank directing deaf persons to the location of a TTY for use at a pay telephone, if there are 4 or more pay phones on the site? (ADA Stds. 4.30.7;4.31.9(3))


## V.  PUBLIC/COMMON USE RESTROOMS

A.    Is each public and employee restroom accessible with at least one large accessible stall/toilet, one accessible lavatory, and one accessible urinal (if urinals are provided)? (ADA Stds. 4.1.3(11;4:22))

B.    Is there adequate room for a person who uses a wheelchair to approach the restroom door from the pull side and pull it open without it hitting the wheelchair—this requires at least

18" of wall space on the latch side of the door? (ADA Stds.4.13.6)

C.  When there is a vestibule into the public or employee restroom that does not have fully automatic doors, is there a 30"x48" clear floor space where one can be outside the swing of a hinged door (i.e.: for out-swinging doors, at least 7'-10" between the exterior door frame and interior door frame) to allow persons who use wheelchairs to proceed through one door without it closing on them and binding the wheelchair as they approach and open the next door? (ADA Stds.4.13.7)

D.  Is each accessible toilet centered 18" from the adjacent side wall, which is the distance that will permit a person with a mobility impairment to use the grab bars? (ADA Stds.4.16.2;4.17.3)

E.  Does each accessible toilet have a horizontal grab bar on the adjacent side wall that is at least 40" long and between 33"-36" above the floor for stabilization and assistance during transfer from a wheelchair? (ADA Stds.4.16.4;4.17.6)

F.  Does each accessible toilet have a horizontal grab bar on the wall behind the toilet that is at least 36" long and between 33"-36" above the floor for stabilization and assistance during transfer from a wheelchair? (ADA Stds.4.16.4;4.17.6)

G.  If the accessible toilet is in a stall, does the stall measure at least 60" wide and 56" deep if it is wall mounted or 59" deep if it is floor mounted to allow persons who use wheelchairs to approach the toilet from a variety of transfer positions (i.e.: diagonal or side approaches)? (ADA Stds.4.17.3)

H.  If the accessible toilet is in a stall, is the stall door positioned diagonally opposite, not directly in front of, the toilet so persons who use wheelchairs may pull fully into the stall without being blocked by the toilet? (ADA Stds.4.17.3)

I.  Is the toilet seat at each accessible toilet between 17"-19" above the floor? (ADA Stds.4.16.3)

J.  If there is a lavatory in the accessible stall, is there 42" between the center of the toilet and the near edge of the adjacent lavatory to permit persons with disabilities to transfer onto or off of the toilet? (ADA Stds.4.17.3)

K.  If there are more than 5 stalls in any restroom, is there one stall in addition to the large accessible stall that is 36" wide, has 2 parallel grab bars at 33"-36" wide and has an out-swinging door for persons with mobility impairments who can walk? (ADA Stds.4.22.4)

L.  Is there at least one lavatory (wash basin) in each public restroom that has each of the following characteristics...?

...a 29" high clearance under the front edge and the top of the bowl no higher than 34" above the floor to allow persons who use wheelchairs to pull under the lavatory and use the faucet hardware? (ADA Stds.4.22.6;4.19.2)

...drains and hot water pipes that are insulated or otherwise configured to protect against contact (ADA Stds.4.19.4)

...a faucet that is easily operable with hardware that is (i.e.: levers, wrist blades, single arm, etc.) usable with one hand, without tight grasping, pinching, or twisting of the wrist, since many persons with disabilities may not have high manual dexterity or use of both hands? (ADA Stds.4.19.5)

M.  Is there an area in each public restroom in which a person who uses a wheelchair can turn around—either a 60" diameter circle or a "T"-shaped turn area? (ADA Stds.4.22.3;4.2.3)

## VI.  INTERIOR SIGNS...

A.  If signs are provided for the following spaces, are the signs mounted on the wall (not the door) to the latch side of the door and centered 60" above the floor so that they can be easily located by persons who are blind or have low vision? (Note: signs may be mounted on the door if they are in addition to the wall mounted signs specified here.) (ADA Stds.4.1.3(66)(a))

All guest rooms

Restaurants, other food service areas, and vending/ice machine areas

Ballrooms and meeting rooms

Public and employee restrooms

Gift shops, newsstands, and other retail shops
Other guest amenities, such as exercise/recreational areas,
   interior pools, business centers, and guest laundry
Mechanical and electrical rooms
Stairways, fire exits, and areas of rescue assistance

B.  Do the wall mounted signs provided for the rooms listed above have Braille and raised letters so that they can be read by persons who are blind or have low vision? (ADA Stds.4.1.3(16)(a);4.30.4)

C.  Are all signs at this lodging facility made without reflective materials, such as, brass, chrome, gold, glass or mirror used as text or background, and have letter and numbers that contrast with the background? (Note: reflective signs are permitted if they are in addition to non-reflective signs.) (ADA Stds.4.1.3(16);4.30.5)


## VII. FIRE ALARM SYSTEM...

A.  If the building has an audible fire alarm system, do each of the following rooms in the hotel have a visual alarm strobe light mounted on the wall at 80" above the floor to alert deaf persons about emergency situations? (ADA Stds.4.1.3(14))
Lobby/front desk
Public corridors
Restaurants, other food service areas, and vending/ice
   machine areas
Ballrooms and meeting rooms
Public and employee restrooms
Gift shops, newsstands, and other retail shops
Accessible guestrooms/suites and additional guestrooms/ suites for use by persons who are deaf or hard of hearing required by the table below...
(Note: In the guestrooms/suites, this requirement may be satisfied by either a fixed strobe unit hardwired to the building-wide fire alarm system, or by a kit which contains a portable strobe unit that can plug into a 110 volt electrical

outlet and is also connected by some means to the building-wide fire alarm system.)

| Number of Rooms In the Hotel | Rooms for Hearing Impaired Guests |
|---|---|
| 1-25 | 1 |
| 26-50 | 2 |
| 51-75 | 3 |
| 76-100 | 4 |
| 101-150 | 5 |
| 151-200 | 6 |
| 201-300 | 7 |
| 301-400 | 8 |
| 401-500 | 9 |
| 501-1000 | 2% of total rooms |
| 1001 and over | 20 plus 1 for each 100 over 1000 |

Other guest amenities, such as exercise/recreational areas, interior pools, business centers, and guest laundry
Other public spaces

## VIII. FOOD SERVICE AREAS...

A.  In each restaurant or other food service establishment on this site, are at least 5% of each type of fixed table or a portion of eating counters (i.e.: where no direct service is provided) accessible, providing a 27" high knee space, at least 19" deep, with table/counter tops at 28"—34" above the floor and split proportionately between smoking and non-smoking areas if provided? (ADA Stds.5.1)

B.   Is there a route to these 5% accessible fixed tables that allows at least 36" clear width and has no abrupt vertical level changes greater than 1/4"? (ADA Stds.5.3;4.3.8)

C.   At each food service establishment on this site where food and/or drink is served for consumption (i.e.: direct counter service) at counters or bars, is there a minimum 60" long portion providing a 27" high knee space, at least 19" deep, with the counter top at 28"-34" above the floor, or service at accessible tables in the same area? (ADA Stds.5.2)

D.   Are the food, drink, condiments and tableware dispensers (juice, coffee, cereal units, condiments, forks, knives, etc. without:

...operational controls or dispensers higher than 54" (i.e.: out of reach to many persons who use wheelchairs)? (ADA Stds.5.5)

...operational controls higher than 46" if they are set back 10"-24" from the edge of the counter (i.e.: out of reach to many persons who use wheelchairs)? (ADA Stds.5.5)

E.   If cafeteria style food service lines are provided, are the tray slides no higher than 34" above the floor and queue lines at least 36" wide (42" minimum if u-turns are required) along the approach? (ADA Stds.5.5)


## IX.   GENERAL GUESTROOM & SUITE ISSUES...

A.   Do entry doors, connecting room doors, and interior doors (except doors to shallow closets) into and within all guest rooms and suites allow 32" clear passage width so persons who use wheelchairs, crutches, and other mobility aids can visit or stay in other rooms? (ADA Std. 9.4)

B.   Do bathroom doors in all guestrooms allow 32" clear passage width so persons who use wheelchairs, crutches, and other mobility aids can visit or stay in other rooms and use the bathroom? (ADA Std. 9.4)

## X.   ACCESSIBLE GUESTROOMS AND SUITES:

A. Does the hotel have the proper number of accessible guestrooms and accessible guestrooms with roll-in shower, based on the Table below? (ADA Std. 9.1.2)

| Standard 9.1.2 | | |
|---|---|---|
| | Column "A" | Column "B" |
| Total Rooms in Facility | Accessible Rooms | Rooms with Roll-in Showers |
| 1 to 25 | 1 | 0 |
| 26 to 50 | 2 | 0 |
| 51 to 75 | 3 | 1 |
| 76 to 100 | 4 | 1 |
| 101 to 150 | 5 | 2 |
| 151 to 200 | 6 | 2 |
| 201 to 300 | 7 | 3 |
| 301 to 400 | 8 | 4 |
| 401 to 500 | 9 | See below* |
| 501 to 1000 | 2% of total rooms | See below* |
| 1001+ | 20+ (1 per 100 | |
| over 1000) | See below* | |

**Note: The number of accessible guest rooms for a given number of rooms in a hotel (left column) is derived by adding together column "A" and column "B."**

*The number of roll-in shower rooms in hotels with more than 400 guestrooms total equals 4+ (1 per 100 rooms over 400).

B.   Are the proper number of guestrooms for persons who are deaf or hard of hearing provided per the Table below? (Note: In addition to 9.1.2 rooms, the rooms required by Table 9.1.3

must have auxiliary visual alarms, notification devices and telephone accommodations.) (ADA Std. 9.1.3)

C.  Are the proper number of the following elements provided for persons who are deaf or hard of hearing in each accessible guestroom/suite, each accessible guestroom/suite required to have a roll-in shower and each additional room for persons who are deaf or hard of hearing required in table 9.1.3: (ADA Std. 9.3.1)

| (Standard 9.1.3) | |
| --- | --- |
| **Number of Rooms In the Hotel** | **Rooms for Hearing Impaired Guests** |
| 1-25 | 1 |
| 26-50 | 2 |
| 51-75 | 3 |
| 76-100 | 4 |
| 101-150 | 5 |
| 151-200 | 6 |
| 201-300 | 7 |
| 301-400 | 8 |
| 401-500 | 9 |
| 501-1000 | 2% of total rooms |

...An electrical outlet within 4' of the telephone jack for TTY use?

...Visual notification device for door knocks and phone calls (Cannot be same strobe as the fire alarm strobe unit)?

...Visual smoke alarm device, if audible smoke alarms are provided in the guestrooms?

...Visual fire alarm (strobe) linked to building-wide fire alarm system, if such system is provided? (Note: In the guestrooms/suites, this requirement may be satisfied by either a fixed strobe unit hardwired to the building-wide fire alarm system, or by a kit which contains a portable strobe unit that can plug into a 110 volt electrical outlet and is also connected by some means to the building-wide fire alarm system.)

D. Have the required accessible guest rooms been distributed among the various types of rooms, such as those listed below, to provide persons with disabilities the same or similar choice as other persons...(ADA Stds. 9.1.4)
...Rooms with One Bed?
...Rooms with Two Beds?
...Connecting Rooms
...Whirlpool Tub Suites?
...Other types of Suites and Guestrooms?


E. Are smoking and non-smoking accessible guestrooms provided based upon the ratio of smoking and non-smoking guestrooms in the facility so persons with disabilities have the same options as everyone else? (ADA Stds. 9.1.4)

F. Do all entry doors to accessible guestrooms and other interior doors (except doors on shallow closets) allow at least 32" of clear passage width to accommodate persons who use wheelchairs, crutches, and walkers? (ADA Stds. 9.2.2(3);4.13.5)

G. Is the door hardware (levers, pulls, panic bars, etc.) on all entry doors to accessible guestrooms and other passage doors within the room usable with one hand, without tight grasping, pinching, or twisting of the wrist, since many persons with disabilities may not have high manual dexterity or use of both hands? (ADA Stds. 9.2.2(3);4.13.9)

H. On the pull side of each door (i.e.: entry door, connecting room door, bathroom doors and other passage doors) in every accessible guestroom/suite, are there at least 18" of clear floor space on the latch side for persons who use wheelchairs, walkers, and other mobility aids to approach and pull open? (ADA Stds. 9.2.2(3);4.13.6)

I. Is the security latch or bolt on the hall door mounted no higher than 48" above the floor so it is within the reach of persons who use wheelchairs and is it operable with one hand, without tight grasping, pinching or twisting of the wrist? (ADA Stds. 9.2.2(3);4.13.9)

J. Regarding maneuvering space around the bed(s) in the accessible guestrooms...

...in accessible guestrooms with one bed, is there at least a 36" wide route on each side of the bed to allow persons who use wheelchairs to transfer onto the bed from either side? (ADA Stds. 9.2.2(1))

...in accessible guestrooms with two beds next to one another, is there a minimum of 36" between the two beds to allow persons who use wheelchairs to transfer onto either bed? (ADA Stds. 9.2.2(1))

...is there clear passage at the foot of each bed that is at least 35" wide so that persons who use wheelchairs can approach and use the accessible features throughout the rooms? (ADA Stds. 9.2.2(2))

K. Are the drapery wands and controls on fixed lamps and HVAC units easily operable with one hand, without tight grasping, pinching or twisting of the wrist, since many persons with disabilities may not have high manual dexterity or use of both hands? (ADA Stds. 9.2.2(5);4.27.4)

L. Are all drapery control wands, fixed lamps and HVAC controls in accessible guestrooms placed within 54" of the floor for side approach or 48" of the floor for forward approach so persons who use wheelchairs can approach and use the controls? (ADA Stds. 9.2.2(5);4.27.3)

M. Are the rod and shelf in the clothes closet or wall mounted unit within 54" of the floor for side approach or 48" of the floor for forward approach so persons who use wheelchairs can approach and use the rod and shelf? (ADA Stds. 9.2.2(4);4.24.3)

N. **Accessible Guestroom Bathroom Issues:**
   1. Do bathroom doors in accessible guestrooms allow at least 32" of clear passage width to accommodate persons who use wheelchairs, walkers and other mobility aids? (ADA Stds.

9.2.2(3);4.13.5)

2. Is the bathroom door hardware (levers, pulls, etc.) easily operable with one hand, without tight grasping, pinching or twisting of the wrist, since many persons with disabilities may not have high manual dexterity or use of both hands? (ADA Stds. 9.2.2(3)(e);4.13.9)

3. Is the accessible toilet in each accessible guestroom bathroom centered 18" from the adjacent side wall, which is the distance that will permit a person with a mobility impairment to use the grab bars? (ADA Stds. 9.2.2(6)(e);4.23.4;4.16.2)

4. Does the toilet in each accessible guestroom bathroom have a horizontal grab bar along the adjacent side wall that is at least 42" long and mounted 33"-36" above the floor for stabilization and assistance during transfer from a wheelchair? (ADA Stds. 9.2.2(6)(e);4.23.4;4.16.2)

5. Does the accessible toilet in each accessible guestroom bathroom have a horizontal grab bar along the wall behind the toilet that is at least 36" long and mounted 33"-36" above the floor for stabilization and assistance during transfer from a wheelchair? (ADA Stds. 9.2.2(6)(e);4.23.4;4.16.2)

6. Is the toilet seat in each accessible toilet room between 17"-19" above the floor? (ADA Stds. 4.16.3

7. Is the lavatory (wash basin) in each accessible guestroom bathroom no more than 34" high with at least 29" high clearance under the front edge to allow persons who use wheelchairs to pull under the lavatory and use the faucet hardware? (ADA Stds. 9.2.2(6)(e);4.23.6;4.19.2)

8. Does the lavatory in each accessible guestroom bathroom have drain and hot water pipes that are insulated or otherwise configured to protect against contact? (ADA Stds. 9.2.2(6)(e);4.23.6;4.19.4)

9. Does the lavatory in each accessible guestroom bathroom have a faucet that is easily operable (i.e.: levers, wrist blades, single arm, etc.) and usable with one hand, without tight grasping, pinching, or twisting of the wrist? (ADA Stds. 9.2.2(6)(e);4.23.6;4.19.5)

10. Is there clear floor space in each accessible guestroom bathroom where persons who use wheelchairs can turn around—either a 60" diameter circle or a "T"-turn area? (ADA Stds. 9.2.2(6)(e))
11. Is there adequate room for a person who uses a wheelchair to approach the bathroom door in each accessible guestroom bathroom from the pull side and pull it open without it hitting the wheelchair? (Note: this requires at least 18" of wall space on the latch side of the door.) (ADA Stds. 9.2.2(3);4.13.6)
12. Are there towel racks or bars placed within 54" of the floor for side approach or 48" of the floor for forward approach so persons who use wheelchairs can approach and use the towel racks? (ADA Stds. 9.2.2(4);4.2.6)
13. Are all of the bathroom floors in the accessible guestrooms slip-resistant so persons who use crutches and walkers do not fall? (ADA Stds. 4.5)

O. **Bathtubs**—Do bathtubs (if provided) in the designated accessible guestrooms, comply with the following:
   1. Are the tub faucet controls positioned between the center of the end wall and the open side of the tub so persons with disabilities may approach and adjust the controls before they transfer onto the tub seat to bathe? (ADA Stds. 4.23.8;4.20.2)
   2. Can the faucet controls and shower diverter be turned on and off easily and are they operable and usable with one hand, without tight grasping, pinching or twisting of the wrist (i.e.: levers, single arm, etc.)? (ADA Stds. 4.23.8;4.20.5)
   3. Is there a transfer tub seat (that can be securely attached to the tub) available for persons who may not be able to stand in the tub to bathe? (ADA Stds. 4.23.8;4.20.3)
   4. Is there an adjustable height hand-held shower wand with at least a 60" long hose provided so persons who bathe from a seated position may wash and rinse with the directional spray? (ADA Stds4.23.8;4.20.6)
   5. Is there a horizontal grab bar at the foot of the tub (by the

controls) that is at least 24" long for stabilization while a person with a disability adjusts the water controls? (ADA Stds. 4.23.8;4.20.4)

6. Is there a horizontal grab bar at the head of the tub that is at least 12" long for stabilization and aid in transfer from a wheelchair to the fixed tub seat? (Note: in tubs with built-in seats at the head of the tub, this grab bar is not required.)

7. Are there two horizontal grab bars (one high/one low) along the side of the tub that are at least 24" long for stabilization and aid in transfer from a wheelchair to the fixed tub seat? (ADA Stds. 4.23.8;4.20.4)

8. Is the gap between the wall and the inside face of each grab bar exactly 1-1/2" to accommodate persons with disabilities who rest their forearms on the bars for stabilization so the arm cannot accidentally pass between the grab bar and wall especially if a fall occurs? (ADA Stds. 4.23.8;4,20,4)

P.   **Roll-in Showers**

1. If there are more than 50 guestrooms/suites in the facility, are the proper number of accessible rooms with roll-in showers provided per table for Standard 9.1.2 below. (Note: accessible rooms with roll-in showers must be provided in addition to standard accessible guestrooms/suites.) (ADA Stds. 9.1.2)

2. If there are more than 50 guestrooms/suites, are all required roll-in showers at least 30" wide by 60" long or 36" wide by 60", so persons who use wheelchairs may transfer to the attached seat or use a shower wheelchair while showering? (ADA Stds. 9.1.2)

3. Do the roll-in showers have a securely fastened folding seat at 17"-19" above the floor onto which persons who use wheelchairs may transfer to shower? (ADA Stds. 4.21.3)

4. Are the faucet controls and shower wand positioned on the wall along the side of the shower seat so they are operable from the folding shower seat or from the shower wheelchair? (ADA Stds. 4.21.5)

5. Is there a horizontal grab bar on the wall alongside the

<table>
<tr><td colspan="3" align="center">Standard 9.1.2</td></tr>
<tr><td rowspan="2">Total Rooms in Facility</td><td>Column "A"</td><td>Column "B"</td></tr>
<tr><td>Accessible Rooms</td><td>Accessible Rooms with Roll-in Showers</td></tr>
<tr><td>1-25</td><td>1</td><td>0</td></tr>
<tr><td>26-50</td><td>2</td><td>0</td></tr>
<tr><td>51-75</td><td>3</td><td>1</td></tr>
<tr><td>76-100</td><td>4</td><td>1</td></tr>
<tr><td>101-150</td><td>5</td><td>2</td></tr>
<tr><td>151-200</td><td>6</td><td>2</td></tr>
<tr><td>201-300</td><td>7</td><td>3</td></tr>
<tr><td>301-400</td><td>8</td><td>4</td></tr>
<tr><td>401-500</td><td>9</td><td>See below*</td></tr>
<tr><td>501-1000</td><td>2% of total rooms</td><td>See below*</td></tr>
<tr><td>1000+</td><td>20+(1 per 100 over 1000)</td><td>See below*</td></tr>
</table>

Note: The total number of accessible guestrooms for a given number of rooms in a hotel (left column) is derived by adding together column "A" and column "B."

---

*The number of roll-in shower rooms in hotels with more than 400 guestrooms total equals 4 + (1 per 100 rooms over 400).

shower seat (but not behind the shower seat) for stabilization and aid in transfer from a wheelchair to the folding shower seat? (ADA Stds. 4.21.2)

6. Is there a horizontal grab bar on the wall opposite the seat for stabilization and aid in maneuvering while in a shower wheelchair? (ADA Stds. 4.21.2)

7. Are the roll-in showers free of doors that would impede wheelchair transfer onto the seat? (ADA Stds. 4.21.8)

8. Are the roll-in showers free of curbs of lips at the shower floor that would impede wheelchair approach and transfer onto the folding shower seat? (ADA Stds. 4.21.7)

9. Do roll-in showers have faucet controls that are easily operable with one hand (i.e.: levers, wrist blades, single arm, etc.) without tight grasping, pinching or twisting of the wrist? (ADA Stds. 4.21.5)
10. Is there an adjustable height shower wand with at least a 60" long hose provided for persons who must shower from a seated position? (ADA Stds. 4.21.6)
11. Is the gap between the wall and the inside face of each grab bar exactly 1-1/2" to accommodate persons with disabilities who rest their forearms on the bars for stabilization so the arm cannot accidentally pass between the grab bar and wall especially if a fall occurs? (ADA Stds. 4.21.2)

**Q.   Other Showers in Accessible Guestrooms/Suites**—Accessible guestrooms/suites that are not required to have roll-in showers may have an accessible bathtub, a small shower or a large shower?

1. **Small Showers—**
   a. Do all small showers, if provided in accessible guestroom/suites, measure exactly 36" wide by 36" deep? (ADA Stds. 4.21.2)
   b. Do they have a fixed or folding seat between 17"-19" above the floor, onto which a person who uses a wheelchair may transfer to shower? (ADA Stds. 4.21.3)
   c. Is there a 36" wide by 48" long clear floor space directly outside the shower for the persons who use wheelchairs to approach and use the shower? (ADA Stds. 4.21.2)
   d. Does the 36"x48" long clear floor space directly outside the shower extend at least 12" past the seat wall of the shower to allow for a seat-to-seat transfer from the wheelchair? (ADA Stds. 4.21.2)
   e. Do all small showers have faucet controls that are easily operable with one hand (i.e.: levers, wrist blades, single arm, etc.) without tight grasping, pinching or twisting of the wrist? (ADA Stds. 4.21.5)
   f. Is there a horizontal grab bar on the wall alongside the shower seat (but not behind the shower seat) for stabi-

lization and to aid in transfer from a wheelchair to the folding shower seat? (ADA Stds. 4.21.4)

g. Is there a horizontal grab bar on the wall opposite the seat? (ADA Stds. 4.21.4)

h. Are the accessible small showers free of doors that would impede wheelchair transfer to the seat? (ADA Stds. 4.21.8)

i. Are all of the small showers free of curbs or lips greater than 1/2" high at the shower floor that would impede wheelchair approach and transfer onto the folding shower seat? (ADA Stds. 4.21.7)

j. Is there an adjustable height shower wand with at least a 60" long hose provided for persons who must shower from a seated position (ADA Stds. 4.21.6)

k. Is the gap between the wall and the inside face of each grab bar exactly 1-1/2" to accommodate persons with disabilities who rest their forearms on the bars for stabilization so the arm cannot accidentally pass between the grab bar and wall especially if a fall occurs? (ADA Stds. 4.21.4)

2. **Large Showers**—Do large showers in accessible guestrooms/ suites, if provided have the following features...

a. Do large showers measure at least 30" wide by 60" deep? (ADA Stds. 4.21.2)

b. Are all large showers free of curbs or lips at the shower floor that would impede wheelchair approach and transfer onto the folding shower seat? (ADA Stds. 4.21.7)

c. Do all large showers have faucet controls that are easily operable with one hand without tight grasping, pinching, or twisting of the wrist (i.e., levers, wrist blades, single arm, etc.)?

d. Is there an adjustable height shower wand with at least a 60" long hose provided for persons who must shower from a seated position? (ADA Stds. 4.21.6)

e. Is there a horizontal grab bar along each of the three shower walls at 33"-36" above the floor? (ADA Stds. 4.21.2)

f.    Is the gap between the wall and the inside face of each grab bar exactly 1-1/2"? (ADA Stds. 4.21.4)

g.    Are the accessible large showers free of doors? (ADA Stds. 4.21.8)


## XI.  OPERATING ISSUES

1. Can persons with disabilities reserve accessible guestrooms/suites in the same ways and on the same terms that other persons can reserve guestrooms/suites?

2. Do all reservations staff (including staff located on-site at the lodging facility and staff located off-site at a reservations center) have ready access to information about the lodging facility's accessible guestrooms/suites (including specific information on types and sizes of accessible showers, bathtubs and other features such as tub seats) for use in making reservations and answering questions?

3. Are accessible guestrooms/suites held for possible use by persons with disabilities until all other rooms in the same price category have been rented?

4. Are rates for accessible guestrooms/suites the same as rates for guestrooms/suites that are not designated accessible?

5. Are accessible features inside and outside the lodging facility maintained in good working order? (For example, repairing cracks in sidewalks on exterior routes; placing portable display racks and potted plants so they do not impede exterior and interior routes; snow removal on exterior routes; replacing damaged or stolen room identification signs; tightening or adjusting accessible toilet seat fasteners, grab bars, handrails and door hardware; battery replacement for TTDs, portable visual smoke alarms and door-knock notification devices; trimming tree branches and shrubs that pose safety hazards for blind persons and persons with low vision.)

6. Are fire-safety information, maximum room rate information, telephone and television information cards, guest ser-

vices guides, restaurant menus, room service menus, and all other printed materials provided for use by guests also available in alternate formats so that blind persons and persons with low vision can read them? (Alternate formats include Braille, large print, and audio recordings)

7. Are accessible guestrooms arranged so that persons who use wheelchairs, crutches and other mobility aids can approach and use beds; bathrooms; closets; heating, air conditioning and drapery controls; lamps and light switches; telephones; computer outlets; mirrors; televisions; balconies; and other room features without moving furniture? (Note: wheelchairs need 36" of clear passage width.)

8. Does the lodging facility allow persons with disabilities to use service animals, without imposing any extra charges or conditions, in guestroom/suites and all public areas of the facility (e.g.: restaurants, bar areas, facility grounds, vans/shuttle buses, and other areas for meeting or recreation)?

9. Does the facility have a reasonable number of TTYs available for use by persons who are deaf or hard of hearing?

10. Is there a TTY available at the front desk so that lodging facility personnel can communicate with persons who are deaf or have speech impairments (e.g.: taking room service orders, answering requests for assistance, etc.)?

11. If the lodging facility has televisions in guestrooms/suites, is a close-captioning decoder provided for use by persons who are deaf or hard of hearing or do televisions include built-in captioning features?

12. Are the hotel staff available to move furniture, and provide and adjust accessible features in guestrooms when features require installation or adjustment to ensure accessibility (e.g.: installing bathtub seats, lowering adjustable shower wands, placing folding seats in transfer showers in the down position, installing auxiliary fire alarm strobes into the building alarm system, activating the television's closed captioning system)?

13. Is the hotel staff trained to offer assistance, upon request, to persons with disabilities who cannot transport their luggage

to/from their guestrooms/suites and who may need assistance in locating guestrooms and hotel amenities?

14. If the facility offers transportation services for guests, is accessible transportation readily available for guests who use wheelchairs and other mobility aids without additional charge?

15. If a portion of the lobby is used for breakfast service, is it approachable and usable by persons with disabilities who cannot climb steps or stairs?

**Source:**
United States Department of Justice, Civil Rights Division. Disability Rights Section.

# Appendix II
# Checklist for Existing Facilities

**THE AMERICANS WITH DISABILITIES ACT
CHECKLIST FOR READILY ACHIEVABLE
BARRIER REMOVAL**

**Source:**
Adaptive Environments Center, Inc.
Barrier Free Environments, Inc.
August, 1995

## PRIORITY 1: ACCESSIBLE APPROACH/ENTRANCE

People with disabilities should be able to arrive on the site, approach the building, and enter as freely as everyone else. At least one route of travel should be safe and accessible for everyone, including people with disabilities.

*Route of Travel*

**Question:** Is there a route of travel that does not require the use of stairs?

**Possible Solutions:** Add a ramp if the route of travel is interrupted by stairs.

Add on alternative route on level ground.

**Question:** Is the route of travel stable, firm and slip-resistant?

**Possible Solutions:** Repair uneven paving.
Fill small bumps and breaks with beveled patches.
Replace gravel with hard top.

**Question:** Is the route at least 36 inches wide?
**Possible Solutions:** Change or move landscaping, furnishing, or other features that narrow the route of travel.
Widen route.

**Question:** Can all objects protruding into the circulation paths be detected by a person with a visual disability using a cane? **In order to be detected** using a cane, an object must be within 27 inches of the ground. Objects hanging or mounted overhead must be higher than 80 inches to provide clear head room. It is not necessary to remove objects that protrude less than 4 inches from the wall.
**Possible Solutions:** Move or remove protruding objects.
Add a cane-detectable base that extends to the ground.
Place a cane-detectable object on the ground underneath as a warning barrier.

**Question:** Do curbs on the route have curb cuts at drives, parking, and drop-offs?
**Possible Solutions:** Install curb cut.
Add small ramp up to curb.

*Ramps*
**Question:** Are the slopes of ramps no greater than 1:12?
**Slope is given as a ratio of the height to**

**the length.** 1:12 means for every 12 inches along the base of the ramp, the height increases one inch. For a 1:12 maximum slope, **at least** one foot of ramp length is needed for each inch of height.

**Possible Solutions:** Lengthen ramp to decrease slope.
Relocate ramp.
If available space is limited, reconfigure ramp to include switchbacks.

**Question:** Do all ramps longer than 6 feet have railings on both sides?

**Possible Solutions:** Add railings.

**Question:** Are railings sturdy, and between 34 and 38 inches high?

**Possible Solutions:** Adjust height of railing if not between 30 and 38 inches.
Secure handrails in fixtures.

**Question:** Is the width between railings or curbs at least 36 inches?

**Possible Solutions:** Relocate the railings.
Widen the ramp.

**Question:** Are ramps non-slip?

**Possible Solutions:** Add non-slip surface material.

**Question:** Is there a 5-foot-long level landing at every 30-foot horizontal length of ramp, at the top and bottom of ramps and at switchbacks?

**Possible Solutions:** Remodel or relocate ramp.

**Question:** Does the ramp rise no more than 30 inches between landings?

**Possible Solutions:** Remodel or relocate ramp.

*Parking and Drop-Off Areas*

| | |
|---|---|
| **Question:** | Are an adequate number of accessible parking spaces available (8 feet wide for car plus 5-foot access aisle)? For guidance in determining the appropriate number to designate, the table below gives the ADAAG requirements for new construction and alterations (for lots with more than 100 spaces, refer to ADAAG): |

| Total Spaces | Accessible |
|---|---|
| 1 to 25 | 1 space |
| 26 to 50 | 2 spaces |
| 51 to 75 | 3 spaces |
| 76 to 100 | 4 spaces |

| | |
|---|---|
| **Possible Solutions:** | Reconfigure a reasonable number of spaces by repainting stripes. |
| **Question:** | Are 8-foot-wide spaces, with minimum 8-foot-wide access aisles, and 98 inches of vertical clearance, available for lift-equipped vans? **At least one of every 8 accessible spaces** must be van-accessible (with a minimum of one van-accessible space in all cases). |
| **Possible Solutions:** | Reconfigure to provide van-accessible space(s). |
| **Question:** | Are the access aisles part of the accessible route to the accessible entrance? |
| **Possible Solutions:** | Add curb ramps.<br>Reconstruct sidewalk. |
| **Question:** | Are the accessible spaces closest to the accessible entrance? |
| **Possible Solutions:** | Reconfigure spaces |

**Question:**                 Are accessible spaces marked with the International Symbol of Accessibility? Are there signs reading "Van Accessible" at van spaces?

**Possible Solutions:**       Add signs, placed so that they are not obstructed by cars.

**Question:**                 Is there an enforcement procedure to ensure that accessible parking is used only by those who need it?

**Possible Solutions:**       Implement a policy to check periodically for violators and report them to the proper authorities.

*Entrance*

**Question:**                 If there are stairs at the main entrance, is there also a ramp or lift, or is there an alternative accessible entrance? **Do not use a service entrance as the accessible entrance** unless there is no other option.

**Possible Solutions:**       If it is not possible to make the main entrance accessible, create a dignified alternate accessible entrance. If parking is provided, make sure there is accessible parking near all accessible entrances.

**Question:**                 Do all inaccessible entrances have signs indicating the location of the nearest accessible entrance?

**Possible Solutions:**       Install signs before inaccessible entrances so that people do not have to retrace the approach.

**Question:**                 Can the alternate accessible entrance be used independently?

**Possible Solutions:**       Eliminate as much as possible the need for assistance—to answer a doorbell, to operate a lift, or to put down a temporary ramp, for example.

**Question:** Does the entrance door have at least 32 inches clear opening (for a double door, at least one 32-inch leaf)?

**Possible Solutions:** Widen the door to 32 inches clear.

If technically feasible, widen to 31-3/8 inches minimum.

Install offset (swing-clear) hinges.

**Question:** Is there at least 18 inches of clear wall space on the pull side of the door, next to the handle? **A person using a wheelchair** or crutches needs this space to get close enough to open the door.

**Possible Solutions:** Remove or relocate furnishings, partitions, or other obstructions.

Move door.

Add power-assisted or automatic door opener.

**Question:** Is the threshold edge 1/4 inch high or less, or if beveled edge, no more than 3/4 inch high?

**Possible Solutions:** If there is a single step with a rise of 6 inches or less, add a short ramp.

If there is a threshold greater than 3/4 inch high, remove it or modify it to be a ramp.

**Question:** If provided, are carpeting or mats a maximum of 1/2 inch high?

**Possible Solutions:** Replace or remove mats.

**Question:** Are edges securely installed to minimize tripping hazards?

**Possible Solutions:** Secure carpeting or mats at edges.

| | |
|---|---|
| **Question:** | Is the door handle no higher than 48 inches and operable with a closed fist? **The "closed fist" test for handles and controls**: Try opening the door or operating the control using only one hand, held in a fist. If it can be done, a person who has limited use of their hands can do it also. |
| **Possible Solutions:** | Lower handle. Replace inaccessible knob with a lever or loop handle. Retrofit with an add-on lever extension. |
| **Question:** | Can doors be opened without too much force (exterior doors reserved; maximum is 5 lbs. for interior doors)? **An inexpensive force meter or a fish scale can be used** to measure the force required to open a door. Attach the hook end to the doorknob or handle. Pull on the ring end until the door opens, and read off the amount of force required. If no force meter or a fish scale is available subjective judgment as to whether the door is easy enough to open can be used. |
| **Possible Solutions:** | Adjust the door closers and oil the hinges. Install power-assisted or automatic door openers. Install lighter doors. |
| **Question:** | If the door has a closer, does it take at least 3 seconds to close? |
| **Possible Solutions:** | Adjust door closer. |

## PRIORITY 2: ACCESS TO GOODS AND SERVICES

Ideally, the layout of the building should allow people with disabilities to obtain materials or services without assistance.

*Horizontal Circulation (ADAAG 4.3)*

**Question:** Does the accessible entrance provide direct access to the main floor, lobby, or elevator?

**Possible Solutions:** Add ramps or lifts.
Make another entrance accessible.

**Question:** Are all public spaces on an accessible route of travel?

**Possible Solutions:** Provide access to all public spaces along an accessible route of travel.

**Question:** Is the accessible route to all public spaces at least 36 inches wide?

**Possible Solutions:** Move furnishings such as tables, chairs, display racks, vending machines, and counters to make more room.

**Question:** Is there a 5-foot circle or a T-shapes space for a person using a wheelchair to reverse direction?

**Possible Solutions:** Move furnishings such as tables, chairs, display racks, vending machines, and counters to make more room.


*Doors (ADAAG 4.13)*

**Question:** Do doors into public spaces have at least a 32-inch clear opening?

**Possible Solutions:** Install offset (swing-clear) hinges.
Widen doors.

**Question:** On the pull side of doors, next to the handle, is there at least 18 inches of clear wall spaces so that a person using a wheelchair or crutches can get near to open the door?

**Possible Solutions:** Reverse the door swing if it is safe to do so.
Move or remove obstructing partitions.

**Question:** Can a door be opened without too much force (5 lb. maximum for interior doors)?

**Possible Solutions:** Adjust or replace closers.

Install lighter doors.

Install power-assisted or automatic door openers.

**Question:** Are door handles 48 inches high or less and operable with a closed fist?

**Possible Solutions:** Lower handles.

Replace inaccessible knobs or latches with lever or loop handles.

Retrofit with add-on levers.

Install power-assisted or automatic door openers.

**Question:** Are all threshold edges 1/4 inch high or less, or if beveled edge, no more than 3/4 inch high?

**Possible Solutions:** If there is a threshold greater than 3/4 inch high, remove it or modify it to be a ramp. If between 1/4 and 3/4 inch high, add bevels to both sides.

*Rooms and Spaces (ADAAG 4.2,4.4,4.5)*

**Question:** Are all aisles and pathways to materials and services at least 36 inches wide?

**Possible Solutions:** Rearrange furnishings and fixtures to clear aisles.

**Question:** Is there a 5-foot circle or T-shaped space for turning a wheelchair completely?

**Possible Solutions:** Rearrange furnishings to clear more room.

**Question:** Is carpeting low-pile, tightly woven, and securely attached along edges?

**Possible Solutions:** Secure edges on all sides.
Replace carpeting.

**Question:** In circulation paths through public areas, are all obstacles cane-detectable (located within 27 inches of the floor or higher than 80 inches, or protruding less than 4 inches from the wall?)

**Possible Solutions:** Remove obstacles.

Install furnishings, planters, or other cane-detectable barriers underneath.

### *Emergency Egress (ADAAG 4.28)*

**Question:** If emergency systems are provided, do they have both flashing lights and audible signals?

**Possible Solutions:** Install visible and audible alarms.

Provide portable devices.

### *Signage for Goods and Services (ADAAG 4.30)*

Different requirements apply to different types of signs.

**Question:** If provided, do signs and room numbers designating permanent rooms and spaces where goods and services are provided comply with the appropriate requirements for such signage?

- Signs mounted with centerline 60 inches from floor.
- Mounted on wall adjacent to latch side of door, or as close as possible.
- Raised characters, sizes between 5/8 and 2 inches high, with contrast (for room numbers, rest rooms, exits).
- Brailled text of the same information.
- If pictogram is used, it must be accompanied by raised characters and Braille.

**Possible Solutions:** Provide signs that have raised letters, Grade II Braille, and that meet all other requirements for permanent room or space signage (See ADAAG 4.1.3(16) and 4.30.)

*Directional and Informational Signage*
The following questions apply to directional and informational signs that fall under Priority 2.

**Question:** If mounted above 80 inches, do they have letters at least 3 inches high, with high contrast, and no-glare finish?

**Possible Solutions:** Review requirements and replace signs are needed, meeting the requirements for character size, contrast, and finish.

**Question:** Do directional and informational signs comply with legibility requirements? (Building directories or temporary signs need not comply.)

**Possible Solutions:** Review requirements and replace signs as needed.

*Controls (ADAAG 4.27)*

**Question:** Are all controls that are available for use by the public (including electrical, mechanical, cabinet, game, and self-service controls) located at an accessible height?

**Reach ranges**: The maximum height for a side reach is 54 inches; for a forward reach, 48 inches. The minimum reachable height is 15 inches for a front approach and 9 inches for a side approach.

**Possible Solutions:** Relocate controls.

**Question:** Are they operable with a closed fist?

**Possible Solutions:** Replace controls.

*Seats, Tables, and Counter (ADAAG 4.2,4.32,7.2)*

**Question:** Are the aisles between fixed seating (other than assembly area seating) at least 36 inches wide?

**Possible Solutions:**    Rearrange chairs or table to provide 36-inch aisles.

**Question:**              Are the spaces for wheelchair seating distributed throughout?

**Possible Solutions:**    Rearrange tables to allow room for wheelchairs in seating areas throughout the area. Remove some fixed seating.

**Question:**              Are the tops of tables or counters between 28 and 34 inches high?

**Possible Solutions:**    Lower part or all of high surface.
Provide auxiliary table or counter.

**Question:**              Are knee spaces at accessible tables at least 27 inches high, 30 inches wide and 19 inches deep?

**Possible Solutions:**    Replace or raise tables.

**Question:**              At each type of cashier counter, is there a portion of the main counter that is no more than 36 inches height?

**Possible Solutions:**    Provide a lower auxiliary counter or folding shelf.
Arrange the counter and surrounding furnishings to create a space to hand items back and forth.

**Question:**              Is there a portion of food-ordering counters that is no more than 36 inches high, or is there space at the side for passing items to customers who have difficulty reaching over a high counter?

**Possible Solutions:**    Lower section of counter.
Arrange the counter and surrounding furnishings to create a space to pass items.

*Vertical Circulation (ADAAG 4.1.3(5), 4.3)*

**Question:** Are there ramps lifts, or elevators to all public levels?

**Possible Solutions:** Install ramps or lifts.
Modify a service elevator.
Relocate goods or services to an accessible area.

**Question:** On each level, if there are stairs between the entrance and/or elevator and essential public areas, is there an accessible alternate route?

**Possible Solutions:** Post clear signs directing people along an accessible route to ramps, lifts, or elevators.

*Stairs (ADAAG 4.9)*

The following questions apply to stairs connecting levels not serviced by an elevator, ramp, or lift.

**Question:** Do treads have a non-slip surface?

**Possible Solutions:** Add non-slip surface to treads.

**Question:** Do stairs have continuous rails on both sides, with extensions beyond the top and bottom stairs?

**Possible Solutions:** Add or replace handrails if possible within existing floor plan.

*Elevators (ADAAG 4.10)*

**Question:** Are both visible and verbal or audible door opening/closing and floor indicators (one tone = up, two tones = down)?

**Possible Solutions:** Install visible and verbal or audible signals.

**Question:** Are the call buttons in the hallway no higher than 42 inches?

**Possible Solutions:**   Lower call buttons.
                          Provide a permanently attached reach stick.

**Question:**             Do the controls inside the cab have raised
                          and Braille lettering?
**Possible Solutions:**   Install raised lettering and Braille next to
                          buttons.

**Question:**             Is there a sign on both door jambs at every
                          floor identifying the floor in raised and
                          Braille letters?
**Possible Solutions:**   Install tactile signs to identify floor num-
                          bers, at a height of 60 inches from floor.

**Question:**             If an emergency intercom is provided, is it
                          usable without voice communication?
**Possible Solutions:**   Modify communication system.

**Question:**             Is the emergency intercom identified by
                          Braille and raised letters?
**Possible Solutions:**   Modify communication system.

**Question:**             Is the emergency intercom identified by
                          Braille and raised letters?
**Possible Solutions:**   Add tactile identification.


*Lifts (ADAAG 4.2,4.11)*
**Question:**             Can the lift be used without assistance? If
                          not, is a call button provided?
**Possible Solutions:**   At each stopping level, post clear instruc-
                          tions for use of the lift.
                          Provide a call button.

**Question:**             Is there at least 30 by 48 inches of clear
                          space for a person in a wheelchair to ap-
                          proach to reach the controls and use the lift?

**Possible Solutions:**   Rearrange furnishings and equipment to clear more space.

**Question:**   Are controls between 15 and 48 inches high up to 54 inches if a side approach is possible)?

**Possible Solutions:**   Move controls.

## PRIORITY 3: USABILITY OF REST ROOMS

When rest rooms are open to the public, they should be accessible to people with disabilities.

### *Getting to the Rest Rooms (ADAAG 4.1)*

**Question:**   If rest rooms are available to the public, is at least one rest room (either one for each sex, or unisex) fully accessible?

**Possible Solutions:**   Reconfigure rest room.
Combine rest rooms to create one unisex accessible

**Question:**   Are there signs at inaccessible rest rooms that give directions to accessible ones?

**Possible Solutions:**   Install accessible signs.

### *Doorways and Passages (ADAAG 4.2,4.13,4.30)*

**Question:**   Is there tactile signage identifying rest rooms?
**Mount signs on the wall,** on the latch side of the odor, complying with the requirements for permanent signage. Avoid using ambiguous symbols in place of text to identify rest rooms.

**Possible Solutions:**   Add access Braille signage, placed to the side of the door, 60 inches to centerline (not on the door itself).

**Question:**            Are pictograms or symbols used to identify rest rooms, and, if used, are raised characters and Braille included below them?

**Possible Solutions:**  If symbols are used, add supplementary verbal signage with raised characters and Braille below pictogram symbol.

**Question:**            Is the doorway at least 32 inches clear?

**Possible Solutions:**  Install offset (swing-clear) hinges.
Widen the doorway.

**Question:**            Are doors equipped with accessible handles (operable with a closed fist), 48 inches high or less?

**Possible Solutions:**  Lower handles
Replace knobs or latches with lever or loop handles.
Add lever extensions.
Install power-assisted or automatic door openers.

**Question:**            Can doors be opened easily (5 lb. maximum force)?

**Possible Solutions:**  Adjust or replace closers.
Install lighter doors.
Install power-assisted or automatic door openers.

**Question:**            Does the entry configuration provide adequate maneuvering space for a person using a wheelchair? **A person in a wheelchair** needs 36 inches of clear width for forward movement, and a 5-foot diameter or T-shaped clear space to make turns. A minimum distance of 48 inches clear of the door swing is needed between the two doors of an entry vestibule.

**Possible Solutions:** Rearrange furnishings such as chairs and trash cans.
Remove inner door if there is a vestibule with two doors.
Move or remove obstructing partitions.

**Question:** Is there a 36-inch wide path to all fixtures?
**Possible Solutions:** Remove obstructions.


*Stalls (ADAAG 4.17)*

**Question:** Is the stall door operable with a closed fist, inside and out?
**Possible Solutions:** Replace inaccessible knobs with lever or loop handles.
Add lever extensions.

**Question:** Is there a wheelchair-accessible stall that has an area of at least 5 feet by 5 feet clear of the door swing, OR is there a stall that is less accessible but that provides greater access than a typical stall (either 36 by 69 inches or 48 by 69 inches)?
**Possible Solutions:** Move or remove partitions.
Reverse the door swing if it is safe to do so.

**Question:** In the accessible stall, are there grab bars behind and on the side wall nearest to the toilet?
**Possible Solutions:** Add grab bars.

**Question:** Is the toilet seat 17 to 18 inches high?
**Possible Solutions:** Add raised seat.


*Lavatories (ADAAG 4.19, 4.24)*

**Question:** Does one lavatory have a 30-inch wide by 48-inch-deep clear space in front? **A maxi-**

|                      | mum of 19 inches of the required depth may be under the lavatory. |
| -------------------- | ------------------------------------------------------------------ |

**Possible Solutions:** Rearrange furnishings.

Replace lavatory.

Remove or alter cabinetry to provide space underneath.

Make sure hot pipes are covered.

Move a partition or wall.

**Question:** Can the faucet be operated with one closed fist?

**Possible Solutions:** Replace with paddle handles.

**Question:** Are soap and other dispensers and hand dryers within reach ranges and usable with one closed fist?

**Possible Solutions:** Lower dispensers.

Replace with or provide additional accessible dispensers.

**Question:** Is the mirror mounted with the bottom edge of the reflecting surface 40 inches high or lower?

**Possible Solutions:** Lower or tilt down mirror.

Add a larger mirror anywhere in the room.

## PRIORY 4: ADDITIONAL ACCESS

*Note that this priority is for items not required for basic access in the first three priorities.*

When amenities such as drinking fountains and public telephones are provided, they should also be accessible to people with disabilities.

*Drinking Fountains (ADAAG 4.15)*

**Question:** Is there at least one fountain with clear floor

|                        | space of at least 30 by 48 inches in front? |
| **Possible Solutions:** | Clear more room by rearranging or removing furnishings. |

**Question:** Is there one fountain with its spout no higher than 36 inches from the ground, and another with a standard height spout (or a single "hi-lo" fountain)?

**Possible Solutions:** Provide cup dispensers for fountains with spouts that are too high.
Provide accessible cooler.

**Question:** Are controls mounted on the front or on the side near the front edge, and operable with one closed fist?

**Possible Solutions:** Replace the controls.

**Question:** Is each water fountain cane-detectable (located within 27 inches of the floor or protruding into the circulation space less than 4 inches from the wall?

**Possible Solutions:** Place a planter or other cane-detectable barrier on each side at floor level.

*Telephones (ADAAG 4.31)*

**Question:** If pay or public use phones are provided, is there clear floor space of at least 30 by 48 inches in front of at least one?

**Possible Solutions:** Move furnishings.
Replace booth with open station.

**Question:** Is the highest operable part of the phone no higher than 48 inches (up to 54 inches if a side approach is possible)?

**Possible Solutions:** Lower telephone.

**Question:**            Does the phone protrude no more than 4 inches into the circulation space?

**Possible Solutions:**  Place a cane-detectable barrier on each side at floor level.

**Question:**            Does the phone have push-button controls?
**Possible Solutions:**  Contact phone company to install push-buttons.

**Question:**            Is the phone hearing-aid compatible?
**Possible Solutions:**  Have phone replaced with a hearing-aid compatible one.

**Question:**            Is the phone adapted with volume control?
**Possible Solutions:**  Have volume control added.

**Question:**            Is the phone with volume control identified with appropriate signage?
**Possible Solutions:**  Add signage.

**Question:**            If there are four or more public phones in the building, is one of the phones equipped with a text telephone (TT or TDD)?
**Possible Solutions:**  Install a text telephone.
Have a portable TT available.
Provided a shelf and outlet next to phone.

**Question:**            Is the location of the text telephone identified by accessible signage bearing the International TDD Symbol?
**Possible Solutions:**  Add signage.

# Building Air Quality Action Plan Verification Checklist

## STEP 1: DESIGNATE AN IAQ MANAGER

(1) An IAQ Manager has been designated.

(2) The IAQ Manager has been educated on the contents of Building Air Quality: A Guide for Building Owners and Facility Managers by reading carefully and possible receiving training in the fundamentals of IAQ.

## STEP 2: DEVELOP AN IAQ PROFILE OF THE BUILDING

**1. Identify and Review Existing Records**

(3) Up-to-date manufacturers' operating instructions and maintenance records for HVAC system components have been reviewed and filed.

(4) Up-to-date schedules and procedures for facility operations and maintenance have been reviewed and filed.

(5) HVAC "as built" blueprints have been updated to indicate current HVAC configuration and filed.

(6) Drawings of tenant build-out and interior building renovations have been updated and filed.

(7) Information on major space use changes (e.g., office space to kitchen or laboratory significant increases or decreases in occupant density) has been updated and filed.

(8) The HVAC system was designed to deliver ____CFM of outside air which translates into ____ CFM of outside air per occupant.

(9) The HVAC system is actually delivering ____ CFM of outside air which translates into ____ CFM of outside air per occupant.

(10) A review of occupant thermal comfort complaints and indoor temperature and relative humidity readings indicates that current peak heating and cooling loads do not exceed HVAC system capacity.

(11) Information on pressure relationships between areas and/or zones within the building has been examined, updated, and filed.

(12) The building's most recent test and balancing report has been filed.
Date of report: ____

(13) Material Safety Data Sheets (MSDS) for products used in the building are requested from suppliers and kept on file.

(14) Documentation of HVAC control system set points and ranges has been reviewed and filed.

(15) The building record (items #3-14) above are revised as needed, particularly with any renovation/construction activities.

**2.    Conduct a Walkthrough to Assess the Current IAQ Situation**

(16) A building walkthrough inspection has been conducted, including both occupied areas and mechanical rooms.

(17) During the walkthrough, a pollutant/source inventory has been competed.
During the walkthrough, IAQ problem indicators have been checked for and noted in a floor plan or comparable drawing, including:

(18) Odors

(19) Dirty or unsanitary conditions

(20) Visible fungal growth or moldy odor

(21) Evident moisture in inappropriate locations (e.g., moisture

on walls, floors or ceilings)
(22) Staining or discoloration of building material(s)
(23) Smoke damage
(24) Presence of hazardous substances
(25) Potential for soil gas entry (e.g., cracks or hole in building surfaces adjacent to ground)
(26) Unusual noises from light fixtures or equipment
(27) Poorly maintained filters
(28) Uneven temperatures
(29) Overcrowding
(30) Personal air cleaner (e.g., ozone generators, portable filtration units) or fans
(31) Inadequate ventilation
(32) Inadequate exhaust air flow
(33) Blocked vents
(34) Other conditions that could impact IAQ, especially risk factors that need regular inspection to prevent IAQ problems from occurring (e.g., drain pans that do not fully drain.)

The condition and operation of the HVAC system has been inspected, including:
(35) Components that need to be repaired, adjusted, cleaned, or replaced have been noted and work orders prepared.
(36) Actual control settings and operating schedules for each air handling unit have been recorded and filed, and checked against the design intent.
(37) Areas with significant sources of containments (e.g., copy rooms, food service areas, printing/photographic areas) are provided with adequate exhaust. Other sources are moved as close to exhaust as possible.


**Notes:**____________________________________________

____________________________________________________

____________________________________________________

____________________________________________________

____________________________________________________

## STEP 3: ADDRESS EXISTING AND POTENTIAL IAQ PROBLEMS

Identified IAQ problems have either been corrected or steps have been taken to control them, including:

(38) source-related IAQ problems
(39) ventilation-related IAQ problems.
(40) Weaknesses have been identified and steps taken to prevent them from becoming problems.

**Notes:**_______________________________________________

_______________________________________________

_______________________________________________

## STEP 4: EDUCATE BUILDING PERSONNEL ABOUT IAQ MANAGEMENT

**(41)** In-house and contractor personnel whose functions could impact IAQ (e.g., housekeeping staff, maintenance contractors) have been identified.

(42) IAQ training or information has been provided to in-house personnel and contractors—especially regarding use of hazardous chemicals. Additional training or information is provided periodically, and plans for continual improvement have been established.

**Notes:**_______________________________________________

_______________________________________________

_______________________________________________

## STEP 5: DEVELOP AND IMPLEMENT A PLAN FOR FACILITY OPERATIONS AND MAINTENANCE

**1. HVAC Operations**

(43) Operating schedules for HVAC equipment, ensuring that the HVAC system is operating during significant occupancy periods, have been written and are updated as needed.

(44) The HVAC operating schedule provides for an adequate flush of the building, with as much outside air as is feasible, prior to occupants' arrival.
**Notes:**_______________________________________________

_______________________________________________

_______________________________________________

2.  **Housekeeping**
(45) All housekeeping equipment and products used in the building are known to the IAQ Manager.
(46) The products used in this building that may produce strong odors, are potential irritants, or may have other IAQ impacts have been determined, and, where possible, have been replaced by products without such impacts.
(47) Housekeeping procedures that detail proper use, storage, and purchase of cleaning materials have been written and are updated as needed.
  The housekeeping staff or contractors have been educated about the IAQ implications, appropriate use, and application of the following to improve IAQ:
(48) Proper cleaning methods
(49) Cleaning schedules
(50) Purchasing
(51) Proper materials storage and use
(52) Proper trash disposal

3.  **HVAC Preventive Maintenance**
(53) A preventive maintenance plan that includes equipment maintenance schedules has been written or computerized and is followed and updated as needed.
  A preventive maintenance plan or contract includes at least the following maintenance areas:
(54) Outside air intakes (inspected for nearby sources of contaminants)
(55) Air distribution dampers (cleared of obstruction and operating properly)

(56) Air filters (pressure drops monitored, replacement or cleaning performed regularly)

(57) Drain pans (inspected and cleaned to ensure proper drainage)

(58) Heating and cooling coils (inspected and cleaned)

(59) Interior of air handling unit (inspected and cleaned as warranted)

(60) Fan motor and belts (inspected)

(61) Air humidification and controls (inspected and regularly cleaned)

(62) Cooling tower (inspected, cleaned, and water treated according to schedule)

(63) Air distribution pathways and VAV boxes (inspected and cleaned as needed).

(64) The preventive maintenance plan and operations manuals are updated when equipment is added, removed, or replaced.

## 4. Unscheduled Maintenance

(65) Procedures for unscheduled maintenance events (e.g., equipment failure) have been written and communicated to building staff. They include:

(66) Building maintenance personnel immediately tell the IAQ Manager that an unscheduled maintenance event has occurred.

(67) Notification to occupants/tenants is provided in a timely manner, addressing how their air quality is being protected.

(68) Necessary remedial action is taken.

**Notes:**_______________________________________________

_______________________________________________

_______________________________________________

_______________________________________________

## General

(69) When new products are purchased, information on potential indoor air contaminant emissions is requested from

product suppliers.
*Note: Emission information may not be readily available for many products at this time; however information that is available should be collected.*
(70) When the services of architects, engineers, contractors, and other professionals are used, IAQ concerns, such as special exhaust needs, are discussed.

## 1.   Remodeling and Renovation
(71) Special procedures to minimize the generation and migration of contaminants or odors to occupied areas of the building are used (or required of contractors).

    The special procedures used in this building are:
(72) The IAQ Manager reviews designs and construction activities for all proposed remodeling and renovation activities prior to their initiation
(73) Work is scheduled during periods of minimum occupancy
(74) Ventilation is provided in order to isolate work areas
(75) Lower-emitting work processes are used (e.g., wet-sanding dry wall)
(76) Specialized cleaning procedures are used (e.g., use of HEPA vacuums)
(77) Filters are changed more frequently, especially after work is completed
(78) Emissions from new furnishings are minimized (e.g., buying lower-emitting products, airing out furnishings before installation, increased amount and duration of ventilation after installation)
(79) Ventilation and distribution equipment are protected.

## 2.   Painting
(80) The exposure to paint vapors is minimized by using low-emitting products, scheduling work during periods of minimum occupancy, or increasing ventilation.

## 3.   Pest Control
(81) Integrated Pest Management procedures are used to the extent possible:

(82) The pest control products being used in the building are known.

(83) Either by written procedures or contract language, it is ensured that all people who use pest control products read and follow all label directions for proper use, mixing, storage and disposal.

(84) Non-chemical pest control strategies are used where possible.

(85) The safest available pest control products that meet that building's needs are purchased or reviewed with pest control contractor.

**4.   Shipping or Receiving**

(86) Vehicle exhaust has been prevented from entering the building (including through air intakes and building openings) by installing barriers to airflow from loading dock areas (e.g., doors, curtains, etc.) and using pressurization.

**Notes:**_______________________________________________

_______________________________________________

_______________________________________________

_______________________________________________

**5.   Smoking**

(87) Smoking is prohibited in all portions of this building, including tenant occupied space.

**OR**

(88) If smoking is permitted in the building, all smoking areas are exhausted directly to the outside, are maintained under negative pressure relative to adjacent space, and are provided with 60 CFM per occupant of make-up air (can be supplied by transfer air).

**Notes:**_______________________________________________

_______________________________________________

_______________________________________________

_______________________________________________

## STEP 7: COMMUNICATE APPROPRIATELY WITH TENANTS/OCCUPANTS ABOUT THEIR ROLE IN MAINTAINING GOOD IAQ

(89) Tenants or occupants are routinely informed about building conditions and policies that may impact IAQ (e.g., practices that attract insects or smoking policy clarifications).

(90) Tenants or occupants are notified about planned major renovation, remodeling, maintenance or pest control activities.

**Notes:**_______________________________________________

_______________________________________________

_______________________________________________

_______________________________________________

_______________________________________________

_______________________________________________

## STEP 8: ESTABLISH PROCEDURES FOR RESPONDING TO IAQ COMPLAINTS

(91) Entries such as IAQ problems are logged into the existing work-order system.

(92) Information is collected from complainants.

(93) Information and records obtained from complainants are kept confidential.

(94) The capability of in-house staff to respond to complaints is assessed.

(95) Appropriate outside sources of assistance are identified.

(96) Feedback is provided in a timely manner to complainant.

(97) Remedial actions are taken.

(98) Remedial actions are followed-up to determine if the action has been effective.

(99) Building staff have been informed of these procedures.

(100) Building occupants and/or tenants have been informed of these procedures and are periodically reminded of how to

locate responsible staff and where to obtain complaint
forms.

**Notes:**_______________________________________

_______________________________________

_______________________________________

_______________________________________

**Source:**

United States Environmental Protection Agency. Air and Radiation (6607J) EPA 401-K-98-001. DHHA (NIOSH) Publication
   No. 98-123. *Building Air Quality Action Plan*, June, 1998.

# Appendix IV
# *Indoor Air Quality Forms*

This section is a series of forms dealing with the various aspects/components of indoor air quality including HVAC, ventilation, complaint forms, interview forms, incident log, etc.

The forms included are:

**IAQ Management Checklist.** This form is used to keep track of the IAQ profile and IAQ management plan.

**Pollutant Pathway Record For IAQ Profiles.** This form is used for identifying areas in which negative or positive pressures should be maintained.

**Zone/Room Record.** This form is used for recording information on a room-by-room basis on the topics of room use, ventilation, and occupant population.

**Ventilation Worksheet.** This form is used in conjunction with the Zone/Room Record when calculating quantities of outdoor air that are being supplied to individual zones or rooms.

**IAQ Complaint Form.** This form is to be filled out by the complainant or by a staff person who receives information from the complainant.

**Incident Log.** This form is used to keep track of each IAQ complaint or problem and how it is handled.

**Occupant Interview.** This form is used for recording the observations of building occupants in relation to their symptoms and conditions in the building.

**Occupant Diary.** This form is used for recording incidents of symptoms and associated observations as they occur.

**Log of Activities and System Operations**. This form is used for recording activities and equipment operating schedules as they occur.

**HVAC Checklist—Short Form**. This form is to be used as a short form for investigating an IAQ problem, or for periodic inspections of the HVAC system. Duplicate these pages for each large air handling unit.

**HVAC Checklist—Long Form**. This form is to be used for the detailed inspections of the HVAC system or as a long form for investigating an IAQ problem. Duplicate these pages for each large air handling unit.

**Pollutant Pathway Form For Investigations**. This form is to be used in conjunction with a floor plan of the building.

**Pollutant and Source Inventory**. This form is to be used as a general checklist of potential indoor and outdoor pollutant sources.

**Chemical Inventory**. This form is used for recording information about chemicals stored or used within the building.

**Hypothesis Form**. This form is used for summarizing what has been learned during the building investigation; a tool to help the investigator collect his or her thoughts.

## IAQ MANAGEMENT CHECKLIST                    Page 1 of 6

Building Name: _________________________________ Date: _______

Address: ___________________________________________________

Completed by (name/title): _________________________________

Use this checklist for inclusion of all the necessary elements in the IAQ profile and IAQ management plan.

| Item | Date begun or completed (as applicable) | Responsible person (name, telephone) | Location ("NA" if the item is not applicable) |
|---|---|---|---|
| **IAQ PROFILE** | | | |
| **Collect and Review Existing Records** | | | |
| HVAC design data, operating instructions and manuals | | | |
| HVAC maintenance and calibration records, testing and balancing reports | | | |
| Inventory of locations where occupancy, equipment, or building use has changed | | | |
| Inventory of complaint locations. | | | |
| **Conduct a Walkthrough Inspection of the Building** | | | |

## IAQ MANAGEMENT CHECKLIST                    Page 2 of 6

| Item | Date begun or completed (as applicable) | Responsible person (name, telephone) | Location ("NA" if the item is not applicable) |
|---|---|---|---|
| List of responsible staff and/or contractors, evidence of training, and job descriptions | | | |
| Identification of areas where positive or negative pressure should be maintained | | | |
| Record of locations that need monitoring or correction | | | |
| **Collect Detailed Information** | | | |
| Inventory of HVAC system components needing repair, adjustment, or replacement | | | |
| Record of control settings and operating schedules | | | |
| Plan showing airflow directions or pressure differentials in significant areas | | | |
| Inventory of significant pollutant sources and their locations | | | |
| Zone/Room Record | | | |
| **IAQ MGMT. PLAN** | | | |

## IAQ MANAGEMENT CHECKLIST  Page 3 of 6

| Item | Date begun or completed (as applicable) | Responsible person (name, telephone) | Location ("NA" if the item is not applicable) |
|---|---|---|---|
| Select IAQ Manager | | | |
| Select IAQ Profile | | | |
| Assign Staff Responsibilities/Train Staff | | | |
| Facilities Operation and Maintenance | | | |
| •confirm that equipment operating schedules are appropriate | | | |
| •confirm appropriate pressure relationships between building usage areas | | | |
| •compare ventilation quantities to design, codes, ASHRAE 62-1989 | | | |
| •schedule equipment inspections per preventive maintenance plan or recommended maintenance schedule | | | |
| •modify and use HVAC Checklist(s); update as equipment is added, removed, or replaced | | | |

## IAQ MANAGEMENT CHECKLIST                    Page 4 of 6

| Item | Date begun or completed (as applicable) | Responsible person (name, telephone) | Location ("NA" if the item is not applicable) |
|---|---|---|---|
| •schedule maintenance activities to avoid creating IAQ problems | | | |
| •review MSDSs for supplies; request additional information as needed | | | |
| •consider using alarms or other devices to signal need for HVAC maintenance (e.g., clogged filters) | | | |
| **Housekeeping** | | | |
| •evaluate cleaning schedules and procedures; modify if necessary | | | |
| •review MSDSs for products in use; buy different products if necessary | | | |
| •confirm proper use and storage of materials | | | |
| •review trash disposal procedures; modify if necessary | | | |
| **Shipping and Receiving** | | | |
| •review loading dock procedures (Note: if air intake is located nearby, take precautions to prevent intake of exhaust fumes.) | | | |

## IAQ MANAGEMENT CHECKLIST                    Page 5 of 6

| Item | Date begun or com-pleted (as applicable) | Responsible person (name, telephone) | Location ("NA" if the item is not applicable) |
|---|---|---|---|
| •check pressure relation-ships around loading dock | | | |
| **Pest Control** | | | |
| •consider adopting Integrated Pest Management methods | | | |
| •obtain and review MSDSs; review handling and storage | | | |
| •review pest control schedules and procedures | | | |
| •review ventilation used during pesticide application | | | |
| **Occupant Relations** | | | |
| •establish health and safety committee or joint tenant/ management IAQ task force | | | |
| •review procedures for responding to complaints; modify if necessary | | | |
| •review lease provisions; modify if necessary | | | |
| **Renovation, Redecorating, Remodeling** | | | |

## IAQ MANAGEMENT CHECKLIST                    Page 6 of 6

| Item | Date begun or completed (as applicable) | Responsible person (name, telephone) | Location ("NA" if the item is not applicable) |
|---|---|---|---|
| •discuss IAQ concerns with architects, engineers, contractors, and other professionals | | | |
| •obtain MSDSs; use materials and procedures that minimize IAQ problems | | | |
| •schedule work to minimize IAQ problems | | | |
| •arrange ventilation to isolate work areas | | | |
| •use installation procedures that minimize emissions from new furnishings | | | |
| **Smoking** | | | |
| •eliminate smoking in the building | | | |
| •if smoking areas are designated, provide adequate ventilation and maintain under negative pressure | | | |
| •work with occupants to develop appropriate non-smoking policies, including implementation of smoking cessation programs | | | |

# POLLUTANT PATHWAY RECORD FOR IAQ PROFILES

This form should be used in combination with a floor plan such as a fire evacuation plan.

Building Name: _________________________________ File Number: _________

Address: _________________________________________________________

Completed by: _____________________________ Title: __________ Date: _______

Building areas that contain contaminant sources (e.g., bathrooms, food preparation areas, smoking lounges, print rooms, and art rooms) should be maintained under negative pressure relative to surrounding areas. Building areas that need to be protected from the infiltration of contaminants (e.g., hallways in multi-family dwellings, computer rooms, and lobbies) should be maintained under positive pressure relative to the outdoors and relative to surrounding areas.

List the building areas in which pressure relationships should be controlled. As the building is inspected, put a Y or N in the "Needs Attention" column to show whether the desired air pressure relationship is present. Mark the floor plan with arrows, plus signs (+) and minus signs (-) to show the airflow patterns observed using chemical smoke or a micromanometer.

Building areas that appear isolated from each other may be connected by airflow passages such as air distribution zones, utility tunnels or chases, party walls, spaces above suspended ceilings (whether or not those spaces are serving as air plenums), elevator shafts, and crawlspaces. If pathways connecting the room to identified pollutant sources (e.g., items of equipment, chemical storage areas, and bathrooms) are observed, record them in the "Comments" column, on the floor plan, or both.

| Building Area (zone, room) | Use | Intended Pressure Positive (+) Negative (-) | | Needs Attention? (Y/N) | Comments |
|---|---|---|---|---|---|
|  |  |  |  |  |  |
|  |  |  |  |  |  |
|  |  |  |  |  |  |
|  |  |  |  |  |  |
|  |  |  |  |  |  |
|  |  |  |  |  |  |

## ZONE/ROOM RECORD

Building Name: _____________________ File Number: _______ Date: _____

Address: _________________ Completed by: __________ Title: _______

This form is to be used differently depending on whether the goal is to prevent or to diagnose IAQ problems. During the development of a profile, this form should be used to record more general information about the entire building; during an investigation, the form should be used to record more detailed information about the complaint area and areas surrounding the complaint area or connected to it by pathways.

Use the last three columns when underventilation is suspected. Use the Ventilation Worksheet to estimate outdoor air quantities. Compare results to the design specifications, applicable building codes, or ventilation guidelines such as ASHRAE 62-1989. Note: For VAV systems, minimum outdoor air under reduced flow conditions must be considered.

| PROFILE AND DIAGNOSIS INFORMATION | | | | | DIAGNOSIS INFORMATION ONLY | | |
|---|---|---|---|---|---|---|---|
| Building Area Zone/ Room | Use[†] | Source of Outdoor Air* | Mechanical Exhaust? (Write "No" or estimate cfm airflow) | Comments | Peak Number of Occupants of Sq. Ft. Floor Area[†] | Total Air Supplied (in cfm)[§] | Outdoor Air Supplied per Person or per 150 Sq. Ft. Area (in cfm) [¶] |
| | | | | | | | |
| | | | | | | | |
| | | | | | | | |
| | | | | | | | |

*Sources might include air handling unit (e.g., AHU-4), operable windows, transfer from corridors
[†]Underline the information in this column if current use or number of occupants is different from design specifications
[§]Mark the information with a P if it comes from the mechanical plans or an M if it becomes from the actual measurements, such as recent test and balance reports
[¶]ASHRAE 62-1989 gives ventilation guidance per 150 sq. ft.

# VENTILATION WORKSHEET

Building Name: ___________________________  File Number: __________

Address:  __________________________________________________________

Completed by (name): ___________________________  Date: __________

This worksheet is designed for use with the **Zone/Room Record**. Formulas are given below for calculating outdoor air quantities using thermal or $CO_2$ information.

The equation for calculating outdoor air quantities **using thermal measurements** is:

$$\text{Outdoor air (in percent)} = \frac{T\,(\text{return air}) \pm T\,(\text{mixed air})}{T\,(\text{return air}) \pm T\,(\text{outdoor air})} \times 100$$

Where: T = temperature in degrees Fahrenheit

The equation for calculating outdoor quantities **using carbon dioxide measurements** is:

$$\text{Outdoor air (in percent)} = \frac{Cs \pm Cs}{Co \pm Cr} \times 100$$

Where:  Cs = ppm of carbon dioxide in the supply air (if measured in a room),
or     Cs = ppm of carbon dioxide in the mixed air (if measured at an air handler)
       Cr = ppm of carbon dioxide in the return air
       Co = ppm of carbon dioxide in the outdoor air

Use the table below to estimate the ventilation rate in any room or zone. Note: ASHRAE 62-1989 generally states ventilation (outdoor air) requirements on an occupancy basis; for a few types of spaces, however, requirements are given on a floor area basis. Therefore, this table provides a process of calculating ventilation (outdoor air) on either an occupancy or floor area basis.

| Zone/ Room | % of Outdoor air | Total Air Supplied to Zone/ Room (cfm) | Peak Occupancy (number of people) Or —Floor Area (square feet) | D = B/C Total Air Supplied Per Person (or per sq. ft. area) | E =(Ax100) x D Outdoor air Supplied Per Person (or per sq. ft. area) |
|---|---|---|---|---|---|
| | A | B | C | D | E |
| | | | | | |
| | | | | | |
| | | | | | |
| | | | | | |
| | | | | | |

## INDOOR AIR QUALITY COMPLAINT FORM

This form can be filled out by the building occupant or by a member of the building staff.

Occupant Name: _________________________________ Date: _________

Department/
Location in Building: _____________________________ Phone: _________

Completed by: ________________ Title: _________ Phone: _________

This form should be used if the complaint may be related to indoor air quality. Indoor air quality problems include concerns with temperature control, ventilation, and air pollutants. Your observations can help to resolve the problem as quickly as possible. Please use the space below to describe the nature of the compliant and any potential causes.

Please list the best time to contact you regarding your IAQ complaint.

_________________________________

So that we can respond promptly, please return this form to:

_________________________________
Contact Person

_________________________________
Room, Building, Mail Code

---

**OFFICE USE ONLY**

File Number: _________ Received By: _______ Date Received: _________

## INCIDENT LOG

Building Name: _________________________ Dates (from): ______ (to): ______

Address: _________________________ Completed by (name): __________

| File Number | Date | Location | Investigation Record (check the forms that were used) | | | | | | | | | Result | Entry By |
|---|---|---|---|---|---|---|---|---|---|---|---|---|---|
| | | | Complaint | Interview | Diary | Log | Zone Record | HVAC Checklist | Pollutant Pathway | Source Inventory | Hypothesis Form | | |
| | | | | | | | | | | | | | |
| | | | | | | | | | | | | | |
| | | | | | | | | | | | | | |
| | | | | | | | | | | | | | |
| | | | | | | | | | | | | | |
| | | | | | | | | | | | | | |
| | | | | | | | | | | | | | |
| | | | | | | | | | | | | | |
| | | | | | | | | | | | | | |
| | | | | | | | | | | | | | |
| | | | | | | | | | | | | | |
| | | | | | | | | | | | | | |
| | | | | | | | | | | | | | |
| | | | | | | | | | | | | | |
| | | | | | | | | | | | | | |
| | | | | | | | | | | | | | |
| | | | | | | | | | | | | | |
| | | | | | | | | | | | | | |

## OCCUPANT INTERVIEW                                    **Page 1 of 2**

Building Name: ______________________ File Number: ___________

Address: _________________________________________________

Occupant Name: __________________ Work Location: __________

Completed by: __________________ Title: _________ Date: _________

### SYMPTOM PATTERNS
What kinds of symptoms or discomfort are being experienced?



Are there other people with similar symptoms or concerns?
                                         Yes _________ No _________


If so, what are their names and locations? ____________________

_________________________________________________________

Are there any health conditions that may cause susceptibility to environmental problems?

- ❏ undergoing chemotherapy or radiation therapy
- ❏ immune system suppressed by disease or other causes
- ❏ chronic cardiovascular disease
- ❏ contact lenses
- ❏ allergies
- ❏ chronic respiratory disease
- ❏ chronic neurological problems

### Timing Patterns
When did the symptoms start?
When are the symptoms generally worse?
Do these symptoms go away? If so, when?

Are there any other events (such as weather events, temperature or humidity changes, or activities in the building) that tend to occur around the same time as the symptoms?

## OCCUPANT INTERVIEW                               **Page 2 of 2**

### SPATIAL PATTERNS
At what place/location are the symptoms or discomfort experienced?

At what place/location in the building is the most time spent?

### ADDITIONAL INFORMATION
Are there any observations about the building conditions that might need attention or might help explain the symptoms (e.g., temperature, humidity, drafts, stagnant air, odors)?

Has medical attention been sought for the symptoms?

Are there any other comments?

## OCCUPANT DIARY

Occupant Name: _____________________ Title: ______ Phone: __________

Location: _______________________________ File Number: ______________

On the form below, record each occasion when a symptom of ill-health or discomfort that may be linked to an environmental condition is experienced in this building.

It is important that the time, date, and location within the building be recorded as accurately as possible. This will help to identify conditions (e.g., equipment operation) that may be associated with the problem. Also describe the severity of the symptoms (e.g., mild, severe) and their duration (the length of time that the symptoms persist). Any other observations should be noted in the "Comments" column. Use additional pages if necessary.

| Time/Date | Location | Symptom | Severity/ Duration | Comments |
|---|---|---|---|---|
|  |  |  |  |  |
|  |  |  |  |  |
|  |  |  |  |  |
|  |  |  |  |  |
|  |  |  |  |  |
|  |  |  |  |  |
|  |  |  |  |  |
|  |  |  |  |  |

## LOG OF ACTIVITIES AND SYSTEM OPERATION

Building Name: _____________ Address: _____________ File No.: _____

Completed by: ________________ Title: _______ Phone: ___________

On the form below, please record observations of the HVAC system operation, maintenance activities and other information that may be helpful in identifying the cause of IAQ complaints in this building. Please report any other observations (e.g., weather, other associated events) that may be also relevant.

Feel free to attach additional pages.

Equipment and activities of particular interest:

Air Handler(s): _______________________________________________

Exhaust Fan(s): _______________________________________________

Other Equipment of Activities: __________________________________

| Date/Time | Day of Week | Equipment Item/Activity | Observations/ Comments |
|---|---|---|---|
|  |  |  |  |
|  |  |  |  |
|  |  |  |  |
|  |  |  |  |
|  |  |  |  |
|  |  |  |  |
|  |  |  |  |

## HVAC CHECKLIST—SHORT FORM                    **Page 1 of 4**

Building Name: ___________ Address: ____________ File No.: _____

Completed by: ________________ Title: _______ Phone: __________

## MECHANICAL ROOM

❑ Clean and dry?___________   ❑ Stored refuse of chemicals? ________
❑ Describe items in need of attention ________________________________

## MAJOR MECHANICAL EQUIPMENT

❑ Preventive maintenance (PM) plan in use? ________________________

### Control System

❑ Type ________________________________________________________
❑ System operation ______________________________________________
❑ Date of last calibration ________________________________________

### Boiler

Rated Btu input ______________   Condition __________________________
❑ Combustion air: is there at least one square inch free area per 2,000
  Btu input? ____________________________________________________
  Fuel or combustion odors _______________________________________

### Cooling

❑ Clean? No leaks or overflow? _______ Slime or algae growth? ______
❑ Eliminator performance ________________________________________
❑ Biocide treatment working? (list type of biocide)___________________
❑ Spill containment plan implemented?_____________________________
❑ Dirt separator working? _______________________________________

### Chillers

❑ Refrigerant leaks? ____________________________________________
❑ Evidence of condensation problems? ______________________________
❑ Waste oil and refrigerant properly stored and disposed of? _______

# HVAC CHECKLIST—SHORT FORM                    Page 2 of 4

Building Name: ___________ Address: ____________ File No.: _____

Completed by: ________________ Title: _______ Phone: ___________

## AIR HANDLING UNIT
❑ Unit identification ___________ Area served____________________

### Outdoor Air Intake, Mixing Plenum, and Dampers
❑ Outdoor air intake location ___________________________________
❑ Nearby contaminant sources? (describe)_________________________
❑ Bird screen in place and unobstructed? _________________________
❑ Design total cfm ___________ outdoor air (O.A.) cfm ___________
   date tested and balanced ______________________________________
❑ Min. % O.A. (damper setting) _________________________________

   Min. cfm O.A. $\dfrac{(\text{total cfm} \times \text{minimum} \% \text{ O.A.})}{100} =$

❑ Current O.A. damper setting
   (date, time, and HVAC operating mode) ___________________
❑ Damper control sequence (describe) ____________________________
❑ Condition of dampers and controls (note date) _________________

### Fans
❑ Control sequence ____________________________________________
❑ Condition (note date) ________________________________________
❑ Indicated temperatures   supply air ______   mixed air ______
                                  return air ______   outdoor air ______
❑ Actual temperatures   supply air ______   mixed air ______
                                  return air ______   outdoor air ______

### Coils
❑ Heating fluid discharge temp ______ $\Delta T$ _______
   Cooling fluid discharge temp ______ $\Delta T$ _______
❑ Controls (describe) __________________________________________
❑ Condition (note date) ________________________________________

### Humidifier
❑ Type _______________________________________________________
   If biocide is used, note type __________________________________
❑ Condition (no overflow, drains trapped, all nozzles working?) _____
❑ No slime, visible growth, or mineral deposits? ________________

## HVAC CHECKLIST — SHORT FORM     Page 3 of 4

Building Name: _____________ Address: _______________ File No.: _____

Completed by: __________________ Title: _______ Phone: ___________

### DISTRIBUTION SYSTEM

| Zone/ Room | System Type | Supply Air | | Return Air | | Power Exhaust | | |
|---|---|---|---|---|---|---|---|---|
| | | ducted/ unducted | cfm | ducted/ unducted | cfm | cfm | control | serves (e.g. toilet) |
| | | | | | | | | |
| | | | | | | | | |
| | | | | | | | | |
| | | | | | | | | |
| | | | | | | | | |
| | | | | | | | | |

**Condition of distribution system and terminal equipment (note locations of problems)**

❑ Adequate access for maintenance? ________________________________

❑ Ducts and coils clean and obstructed? ___________________________

❑ Air paths unobstructed? supply ______ return ______
                 transfer ____ exhaust _____ make-up _____

❑ Note locations of blocked air paths, diffusers, or grilles __________

❑ Any unintentional openings into plenums? _______________________

❑ Controls operating properly? __________________________________

❑ Air volume correct? _________________________________________

❑ Drain pans clean? Any visible growth or odors? ________________

**Filters**

| Location | Type/Rating | Size | Date Last Changed | Condition (give date) |
|---|---|---|---|---|
| | | | | |
| | | | | |
| | | | | |
| | | | | |
| | | | | |

## HVAC CHECKLIST—SHORT FORM                    **Page 4 of 4**

Building Name: _____________ Address: _____________ File No.: _____

Completed by: _________________ Title: _______ Phone: ___________

### OCCUPIED SPACE

**Thermostat types** _________________________________________

| Zone/ Room | Thermostat Location | Thermostat Control (radiator, AHU-3) | Setpoints | | Measured Temperature | Day/Time |
|---|---|---|---|---|---|---|
| | | | Summer | Winter | | |
| | | | | | | |
| | | | | | | |
| | | | | | | |
| | | | | | | |

**Humidistat/Dehumidistat types** _________________________________

| Zone/ Room | Humidistat/ Dehumidistat Location | Control Areas | Setpoints (%RH) | Measured Temperature | Day/Time |
|---|---|---|---|---|---|
| | | | | | |
| | | | | | |
| | | | | | |
| | | | | | |

❏Potential problems (note location) _______________________________

❏Thermal comfort or air circulation problems (drafts, obstructed airflow, stagnant air, overcrowding, poor thermostat location) _______________

_____________________________________________________________

_____________________________________________________________

❏Malfunctioning equipment _______________________________________

❏Major sources of odors or contaminants (e.g., poor sanitation, incompatible uses of space) _________________________________________

_____________________________________________________________

_____________________________________________________________

## HVAC CHECKLIST—LONG FORM                    **Page 1 of 11**

Building Name: _____________ Address: _______________ File No.: _____

Completed by: _________________ Title: _______ Phone: __________

| Component | OK | Needs Attention | Not Applicable | Comments |
|---|---|---|---|---|
| **Outside Air Intake** | | | | |
| Location _____________ <br> _____________ | | | | |
| Open during occupied hours? | | | | |
| Unobstructed? | | | | |
| Standing water, bird droppings in vicinity? | | | | |
| Carryover of exhaust heat? | | | | |
| Cooling tower within 25 feet? | | | | |
| Exhaust outlet within 25 feet? | | | | |
| Trash compactor within 25 feet? | | | | |
| Near parking facility, <br> busy road, <br> loading dock? | | | | |
| **Bird Screen** | | | | |
| Unobstructed? | | | | |
| General condition? | | | | |
| Size of mesh (1/2" minimum) | | | | |
| **Outside Air Dampers** | | | | |
| Operation acceptable? | | | | |
| Seal when closed? | | | | |
| Actuators operation? | | | | |

## HVAC CHECKLIST—LONG FORM  Page 2 of 11

Building: _________________________________  File Number: _____

Completed by: _____________ Title: _________ Date Checked: _______

| Component | OK | Needs Attention | Not Applicable | Comments |
|---|---|---|---|---|
| **Outdoor Air (O.A.) Quantity** (*Check against applicable codes and ASHRAE 62-1989.*) | | | | |
| Minimum % O.A. ___________ | | | | |
| Measured % O. A. __________ *Note day, time, HVAC operating mode under "Comments"* | | | | |
| Maximum % O.A. __________ | | | | |
| Is minimum O.A. a separate damper? | | | | |
| For VAV systems: is O.A. increased as total system air flow is reduced? | | | | |
| | | | | |
| **Mixing Plenum** | | | | |
| Clean? | | | | |
| Floor drain trapped? | | | | |
| Airtightness | | | | |
| •of outside air dampers | | | | |
| • of return air damper | | | | |
| • of exhaust air dampers | | | | |
| All damper motors connected? | | | | |
| All damper motors operational? | | | | |
| Air mixers or opposed blades? | | | | |
| Mixed air temperature control setting ___________°F | | | | |
| Freeze stat setting ________°F | | | | |
| Is mixing plenum under negative pressure? *Note: If it is under positive pressure, outdoor air may not be entering.* | | | | |

# HVAC CHECKLIST—LONG FORM                    Page 3 of 11

Building: _________________________________ File Number: ________

Completed by: _______________ Title: _______ Date Checked: _______

| Component | OK | Needs Attention | Not Applicable | Comments |
|---|---|---|---|---|
| **Filters** | | | | |
| Type ________ | | | | |
| Complete coverage? (i.e., no bypassing) | | | | |
| Correct pressure drip? (*Compare to manufacturer's recommendations.*) | | | | |
| Contaminants visible? | | | | |
| Odor noticeable? | | | | |
| | | | | |
| **Spray Humidifiers or Air Washers** | | | | |
| Humidifier type | | | | |
| All nozzles working? | | | | |
| Complete coil coverage? | | | | |
| Pans clean, no overflow? | | | | |
| Drains trapped? | | | | |
| Biocide treatment working? *Note: Is MSDS on file?* _________ | | | | |
| Spill contaminant system in place? | | | | |
| | | | | |
| **Face and Bypass Dampers** | | | | |
| Damper operation correct? | | | | |
| Damper motors operational? | | | | |
| | | | | |
| **Cooling Coil** | | | | |
| Inspection access? | | | | |
| Clean? | | | | |
| Supply water temp. _____°F | | | | |
| Water carryover? | | | | |
| Any indication of condensation problems? | | | | |

# HVAC CHECKLIST—LONG FORM                    **Page 4 of 11**

Building: _______________________________ File Number: _________

Completed by: ______________ Title: ______ Date Checked: ________

| Component | OK | Needs Attention | Not Applicable | Comments |
|---|---|---|---|---|
| **Condensate Drip Pans** | | | | |
| Accessible to inspect and clean? | | | | |
| Clean, no residue? | | | | |
| No standing water, no leaks? | | | | |
| Noticeable odor? | | | | |
| Visible growth (e.g., slime)? | | | | |
| Drains and traps clear, working? | | | | |
| Trapped to air gap? | | | | |
| Water overflow? | | | | |
| | | | | |
| **Mist Eliminators** | | | | |
| Clean, straight, no carryover? | | | | |
| | | | | |
| | | | | |
| | | | | |
| **Supply Fan Chambers** | | | | |
| Clean? | | | | |
| No trash or storage? | | | | |
| Floor drain traps are wet or sealed? | | | | |
| No air leaks? | | | | |
| Doors close tightly? | | | | |
| | | | | |
| | | | | |
| **Supply Fans** | | | | |
| Location ___________ | | | | |
| Fan blades clean? | | | | |
| Belt guards installed? | | | | |
| Proper felt tension? | | | | |
| Excess vibration? | | | | |
| Corrosion problems? | | | | |
| Controls operational, calibrated? | | | | |
| Control sequence conforms to design/specifications? (describe changes) | | | | |

## HVAC CHECKLIST—LONG FORM                    Page 5 of 11

Building: _________________________________ File Number: _______

Completed by: ________________ Title: _____ Date Checked: _______

| Component | OK | Needs Attention | Not Applicable | Comments |
|---|---|---|---|---|
| No pneumatic leaks? | | | | |
| | | | | |
| | | | | |
| | | | | |
| **Heating Coil** | | | | |
| Inspection access? | | | | |
| Clean? | | | | |
| Control sequence conforms to design/specifications? (describe changes) | | | | |
| Supply water temp. _____°F | | | | |
| Discharge thermostat? (air temp. setting _______°F | | | | |
| | | | | |
| **Reheat Coils** | | | | |
| Clean? | | | | |
| Obstructed? | | | | |
| Operational? | | | | |
| | | | | |
| **Steam Humidifier** | | | | |
| Humidifier type _________ | | | | |
| Treated builder water? | | | | |
| Standing water? | | | | |
| Visible growth | | | | |
| Mineral deposits? | | | | |
| Control setpoint _______°F | | | | |
| High limit setpoint _____°F | | | | |
| Duct liner within 12 feet? (If so, check for dirt, mold growth.) | | | | |
| | | | | |
| **Supply Ductwork** | | | | |
| Clean? | | | | |
| Sealed, no leaks, tight connections | | | | |

## HVAC CHECKLIST—LONG FORM                    **Page 6 of 11**

Building: _________________________________ File Number: _______

Completed by: _______________ Title: ______ Date Checked: ________

| Component | OK | Needs Attention | Not Applicable | Comments |
|---|---|---|---|---|
| Fire dampers open? | | | | |
| Access doors closed? | | | | |
| Lined ducts? | | | | |
| Flex duct connected, no tears? | | | | |
| Light troffer supply? | | | | |
| Balanced within 3-5 years? | | | | |
| Balanced after recent renovations? | | | | |
| Short circuiting or other air distribution problems? Note location(s) _________________ _________________ | | | | |
| | | | | |

**Pressurized Ceiling Supply Plenum**

| Component | OK | Needs Attention | Not Applicable | Comments |
|---|---|---|---|---|
| No unintentional openings? | | | | |
| All ceiling tiles in place? | | | | |
| Barrier paper correctly placed and in good condition? | | | | |
| Proper layout for air distribution? | | | | |
| Supply diffusers open? | | | | |
| Supply diffusers balanced? | | | | |
| Balancing capability? | | | | |
| Noticeable flow of air? | | | | |
| Short circuiting or other air distribution problems? *Note locations(s) in "Comments"* | | | | |
| | | | | |

**Terminal Equipment (supply)**

| Component | OK | Needs Attention | Not Applicable | Comments |
|---|---|---|---|---|
| Housing interiors clean and unobstructed? | | | | |
| Controls working? | | | | |
| Delivering rated volume? | | | | |
| Balanced within 3-5 years? | | | | |
| Filters in place? | | | | |
| Condensate pans clean, drain freely? | | | | |

## HVAC CHECKLIST—LONG FORM      **Page 7 of 11**

Building: _________________________________ File Number: _______

Completed by: _______________ Title: _____ Date Checked: _______

| Component | OK | Needs Attention | Not Applicable | Comments |
|---|---|---|---|---|
| **VAV Box** | | | | |
| Minimum stops ______% | | | | |
| Minimum outside air ___% *(From page 2 of this form)* | | | | |
| Minimum airflow ______ cfm | | | | |
| Minimum outside air ___ cfm | | | | |
| Supply setpoint ___°F (summer) ___°F (winter) | | | | |
| | | | | |
| **Humidity Sensor** | | | | |
| Humidistat setpoints ______% RH | | | | |
| Actual RH _________% | | | | |
| | | | | |
| **Room Partitions** | | | | |
| Gap allowing airflow at top? | | | | |
| Gap allowing airflow at bottom? | | | | |
| Supply and return each room? | | | | |
| | | | | |
| **Stairwells** | | | | |
| Doors close and latch? | | | | |
| No openings allowing uncontrolled airflow? | | | | |
| Clean, dry? | | | | |
| No noticeable odors? | | | | |
| | | | | |
| **Return Air Plenum** | | | | |
| Tiles in place? | | | | |
| No unintentional openings? | | | | |
| Return grilles? | | | | |
| Balancing capability? | | | | |
| Noticeable flow of air? | | | | |
| Transfer grilles? | | | | |
| Fire dampers open? | | | | |

## HVAC CHECKLIST—LONG FORM        **Page 8 of 11**

Building: _________________________________ File Number: _______

Completed by: _______________ Title: _____ Date Checked: _______

| Component | OK | Needs Attention | Not Applicable | Comments |
|---|---|---|---|---|
| **Ducted Returns** | | | | |
| Balanced within 3-5 years? | | | | |
| Unobstructed grills? | | | | |
| Unobstructed return air path? | | | | |
| | | | | |
| **Return Fan Chambers** | | | | |
| Clean and no trash or storage? | | | | |
| No standing water? | | | | |
| Floor drain traps are wet or sealed? | | | | |
| No air leaks? | | | | |
| Doors close tightly, kept closed? | | | | |
| | | | | |
| **Return Fans** | | | | |
| Location _____________ | | | | |
| Fan blades clean? | | | | |
| Belt guards installed? | | | | |
| Proper belt tension? | | | | |
| Excess vibration? | | | | |
| Corrosion problems? | | | | |
| Controls working, calibrated? | | | | |
| Control sequence conforms to design/specifications? (describe changes) | | | | |
| | | | | |
| **Exhaust Fans** | | | | |
| Central? | | | | |
| Distributed (locations) _________ _____________ | | | | |
| Operational? | | | | |
| Controls operational? | | | | |
| Toilet exhaust only? | | | | |
| Gravity relief? | | | | |
| Total powered exhaust _____cfm | | | | |
| Make-up air sufficient? | | | | |

## HVAC CHECKLIST—LONG FORM                    **Page 9 of 11**

Building: _________________________________ File Number: _______

Completed by: _______________ Title: _____ Date Checked: _______

| Component | OK | Needs Attention | Not Applicable | Comments |
|---|---|---|---|---|
| **Toilet Exhausts** | | | | |
| Fans working occupied hours? | | | | |
| Registers open, clear? | | | | |
| Make-up air path adequate? | | | | |
| Volume according to code? | | | | |
| Floor drain traps wet or sealable? | | | | |
| Bathrooms run slightly negative relative to building? | | | | |
| | | | | |
| **Smoking Lounge Exhaust** | | | | |
| Room runs negative relative to building? | | | | |
| | | | | |
| **Print Room Exhaust** | | | | |
| Room runs negative relative to building? | | | | |
| | | | | |
| **Garage Ventilation** | | | | |
| Operates according to codes? | | | | |
| Fans, controls, dampers all operate? | | | | |
| Garage slightly negative relative to building? | | | | |
| Doors to building close tightly? | | | | |
| Vestibule entrance to building from garage? | | | | |
| | | | | |
| **Mechanical Rooms** | | | | |
| General condition? | | | | |
| Controls operational? | | | | |
| Pneumatic controls: | | | | |
| •compressor operational? | | | | |
| •air dryer operational? | | | | |
| Electric controls? Operational? | | | | |

# HVAC CHECKLIST—LONG FORM    Page 10 of 11

Building: _________________________________ File Number: _______

Completed by: ________________ Title: _____ Date Checked: ________

| Component | OK | Needs Attention | Not Applicable | Comments |
|---|---|---|---|---|
| EMS (Energy Management Systems) or DDC (Direct Digital Control): | | | | |
| •operator on site? | | | | |
| •controlled off-site? | | | | |
| •area fans cycled "off" while building is occupied? | | | | |
| •is chiller reset to shed load? | | | | |
| | | | | |
| **Preventive Maintenance** | | | | |
| Spare parts inventoried? | | | | |
| Spare air filters? | | | | |
| Control drawing posted? | | | | |
| PM (Preventive Maintenance) schedule available? | | | | |
| PM followed? | | | | |
| | | | | |
| **Boilers** | | | | |
| Flues, breeching tight? | | | | |
| Purge cycle working? | | | | |
| Door gaskets tight? | | | | |
| Fuel system tight, no leaks? | | | | |
| Combustion air: at least 1 square inch free area per 2000 Btu input? | | | | |
| | | | | |
| **Cooling Tower** | | | | |
| Sump clean? | | | | |
| No leaks, no overflow? | | | | |
| Eliminators working, no carryover? | | | | |
| No slime or algae? | | | | |
| Biocide treatment working? | | | | |
| Dirt separator working? | | | | |

## HVAC CHECKLIST—LONG FORM     **Page 11 of 11**

Building: ________________________________ File Number: _______

Completed by: _______________ Title: _____ Date Checked: _______

| Component | OK | Needs Attention | Not Applicable | Comments |
|---|---|---|---|---|
| **Chillers** | | | | |
| No refrigerant leaks? | | | | |
| Purge cycle normal? | | | | |
| Waste oil, refrigerant properly disposed of and spare refrigerant properly stored? | | | | |
| Condensation problems? | | | | |
| | | | | |
| | | | | |
| | | | | |
| | | | | |
| | | | | |
| | | | | |
| | | | | |
| | | | | |
| | | | | |
| | | | | |
| | | | | |
| | | | | |
| | | | | |
| | | | | |
| | | | | |
| | | | | |
| | | | | |
| | | | | |
| | | | | |
| | | | | |
| | | | | |
| | | | | |
| | | | | |
| | | | | |

# HVAC CHECKLIST—LONG FORM                    **Page __ of __**

Building: _________________________________ File Number: _______

Completed by: _______________ Title: _____ Date Checked: _______

| Component | OK | Needs Attention | Not Applicable | Comments |
|---|---|---|---|---|
|  |  |  |  |  |
|  |  |  |  |  |
|  |  |  |  |  |
|  |  |  |  |  |
|  |  |  |  |  |
|  |  |  |  |  |
|  |  |  |  |  |
|  |  |  |  |  |
|  |  |  |  |  |
|  |  |  |  |  |
|  |  |  |  |  |
|  |  |  |  |  |
|  |  |  |  |  |
|  |  |  |  |  |
|  |  |  |  |  |
|  |  |  |  |  |
|  |  |  |  |  |
|  |  |  |  |  |
|  |  |  |  |  |
|  |  |  |  |  |
|  |  |  |  |  |
|  |  |  |  |  |
|  |  |  |  |  |
|  |  |  |  |  |
|  |  |  |  |  |
|  |  |  |  |  |
|  |  |  |  |  |
|  |  |  |  |  |
|  |  |  |  |  |
|  |  |  |  |  |

## HVAC CHECKLIST—LONG FORM                     Page __ of __

Building: _________________________________ File Number: _______

Completed by: _______________ Title: _____ Date Checked: _______

| Component | OK | Needs Attention | Not Applicable | Comments |
|---|---|---|---|---|
| | | | | |
| | | | | |
| | | | | |
| | | | | |
| | | | | |
| | | | | |
| | | | | |
| | | | | |
| | | | | |
| | | | | |
| | | | | |
| | | | | |
| | | | | |
| | | | | |
| | | | | |
| | | | | |
| | | | | |
| | | | | |
| | | | | |
| | | | | |
| | | | | |
| | | | | |
| | | | | |
| | | | | |
| | | | | |
| | | | | |
| | | | | |
| | | | | |
| | | | | |
| | | | | |
| | | | | |
| | | | | |
| | | | | |

## POLLUTANT PATHWAY FORM FOR INVESTIGATIONS

Building: _________________________________ File Number: ___________

Completed by: _________________ Title: ______ Date Checked: _______

This form should be used in combination with a floor plan such as a fire evacuation plan.

Building areas that appear isolated from each other may be connected by airflow passages such as air distribution zones, utility tunnels or chases, party walls, spaces above suspended ceilings (whether or not those spaces are serving as air plenums), elevator shafts, and crawl spaces.

Describe the complaint area in the space below and mark it on your floor plan. Then list rooms or zones connected to the compliant area by airflow pathways. Use the form to record the direction of air flow between the complaint area and the connected rooms/zones, including the date and time. (Airflow patterns generally change over time). Mark the floor plan with arrows or plus (+) and minus (-) signs to map out the airflow patterns observed, using chemical smoke or a micromanometer. The "Comments" column can be used to note pollutant sources that merit further attention.

Rooms or zones included in the complaint area: ___________________

_______________________________________________________________

| Rooms or Zones Connected to the Complaint Area By Pathways | Use | Pressure Relative to Complaint Area | | Comments (e.g., potential pollutant sources) |
|---|---|---|---|---|
| | | +/- | date/time | |
| | | | | |
| | | | | |
| | | | | |
| | | | | |
| | | | | |
| | | | | |
| | | | | |

## POLLUTANT AND SOURCE INVENTORY          Page 1 of 6

Building: _________________________________ File Number: ________

Completed by: ________________ Title: _____ Date Checked: _______

Using the list of potential source categories below, record any indications of contamination or suspected pollutants that may require further investigation or treatment. Sources of contamination may be constant or intermittent or may be linked to single, unrepeated events. For intermittent sources, indicate the time of peak activity or contaminant production, including correlations with weather (e.g., wind direction).

| Source Category | Checked | Needs Attention | Location | Comments |
|---|---|---|---|---|
| **SOURCES OUTSIDE BUILDING** | | | | |
| **Contaminated Outdoor Air** | | | | |
| Pollen, dust | | | | |
| Industrial contaminants | | | | |
| General vehicular contaminants | | | | |
| | | | | |
| **Emissions from Nearby Sources** | | | | |
| Vehicles exhaust (parking areas, loading docks, roads) | | | | |
| Dumpsters | | | | |
| Re-entrained exhaust | | | | |
| Debris near outside air intake | | | | |
| | | | | |
| **Soil Gas** | | | | |
| Radon | | | | |
| Leaking underground tanks | | | | |
| Sewage smells | | | | |
| Pesticides | | | | |

## POLLUTANT AND SOURCE INVENTORY                    Page 2 of 6

Building: _______________________________ File Number: ________

Completed by: _______________ Title: _______Date Checked: _______

Using the list of potential source categories below, record any indications of contamination or suspected pollutants that may require further investigation or treatment. Sources of contamination may be constant or intermittent or may be linked to single, unrepeated events. For intermittent sources, indicate the time of peak activity or contaminant production, including correlations with weather (e.g., wind direction).

| Source Category | Checked | Needs Attention | Location | Comments |
|---|---|---|---|---|
| **Moisture or Standing Water** | | | | |
| Rooftop | | | | |
| Crawlspace | | | | |
| | | | | |
| | | | | |
| **EQUIPMENT** | | | | |
| **HVAC System Equipment** | | | | |
| Combustion gases | | | | |
| Dust, dirt, or microbial growth in ducts | | | | |
| Microbial growth in drip pans, chillers, humidifiers | | | | |
| Leaks of treated boiler water | | | | |
| | | | | |
| | | | | |
| **Non HVAC System Equipment** | | | | |
| Office Equipment | | | | |
| Supplies for Equipment | | | | |
| Laboratory Equipment | | | | |

## POLLUTANT AND SOURCE INVENTORY          Page 3 of 6

Building: _________________________________ File Number: _________

Completed by: _________________ Title: ______ Date Checked: _______

Using the list of potential source categories below, record any indications of contamination or suspected pollutants that may require further investigation or treatment. Sources of contamination may be constant or intermittent or may be linked to single, unrepeated events. For intermittent sources, indicate the time of peak activity or contaminant production, including correlations with weather (e.g., wind direction).

| Source Category | Checked | Attention | Location | Needs Comments |
|---|---|---|---|---|
| **HUMAN ACTIVITIES** | | | | |
| **Personal Activities** | | | | |
| Smoking | | | | |
| Cosmetics (odors) | | | | |
| | | | | |
| | | | | |
| **Housekeeping Activities** | | | | |
| Cleaning materials | | | | |
| Cleaning procedures (e.g., dust from sweeping, vacuuming) | | | | |
| Stored supplies | | | | |
| Stored refuse | | | | |
| | | | | |
| | | | | |
| **Maintenance Activities** | | | | |
| Use of materials with volatile compounds (e.g., paint, caulk, adhesives) | | | | |
| Stored supplies with volatile compounds | | | | |
| Use of pesticides | | | | |

## POLLUTANT AND SOURCE INVENTORY          Page 4 of 6

Building: ___________________________ File Number: _______

Completed by: _______________ Title: _____ Date Checked: _______

Using the list of potential source categories below, record any indications of contamination or suspected pollutants that may require further investigation or treatment. Sources of contamination may be constant or intermittent or may be linked to single, unrepeated events. For intermittent sources, indicate the time of peak activity or contaminant production, including correlations with weather (e.g., wind direction).

| Source Category | Checked | Attention | Location | Needs Comments |
|---|---|---|---|---|
| **BUILDING COMPONENTS FURNISHINGS** | | | | |
| **Locations Associated with Dust or Fibers** | | | | |
| Dust-catching area (e.g., open shelving) | | | | |
| Deteriorated furnishings | | | | |
| Asbestos-containing materials | | | | |
| | | | | |
| | | | | |
| | | | | |
| | | | | |
| | | | | |
| | | | | |
| | | | | |
| **Unsanitary Conditions/Water Damage** | | | | |
| Microbial growth in or on soiled or water-damaged furnishings | | | | |
| | | | | |
| | | | | |
| | | | | |

## POLLUTANT AND SOURCE INVENTORY          Page 5 of 6

Building: _________________________________ File Number: _______

Completed by: _________________ Title: _____ Date Checked: ______

Using the list of potential source categories below, record any indications of contamination or suspected pollutants that may require further investigation or treatments. Sources of contamination may be constant or intermittent or may be linked to single, unrepeated events. For intermittent sources, indicate the time of peak activity or contaminant production, including correlations with weather (e.g., wind direction).

| Source Category | Checked | Needs Attention | Location | Comments |
|---|---|---|---|---|
| **Chemicals Released From Building Components or Furnishings** | | | | |
| Volatile compounds | | | | |
| | | | | |
| | | | | |
| | | | | |
| | | | | |
| | | | | |
| | | | | |
| | | | | |
| **OTHER SOURCES** | | | | |
| **Accidental Events** | | | | |
| Spills (e.g., water, chemicals, beverages) | | | | |
| Water leaks or flooding | | | | |
| Fire damage | | | | |
| | | | | |
| | | | | |
| | | | | |
| | | | | |

## POLLUTANT AND SOURCE INVENTORY            Page 6 of 6

Building: _________________________________ File Number: _________

Completed by: _________________ Title: _____ Date Checked: _______

Using the list of potential source categories below, record any indications of contamination or suspected pollutants that may require further investigation or treatments. Sources of contamination may be constant or intermittent or may be linked to single, unrepeated events. For intermittent sources, indicate the time of peak activity or contaminant production, including correlations with weather (e.g., wind direction).

| Source Category | Checked | Needs Attention | Location | Comments |
|---|---|---|---|---|
| **Special Use/Mixed Use Areas** | | | | |
| Smoking lounges | | | | |
| Food preparation areas | | | | |
| Underground or attached parking garages | | | | |
| Laboratories | | | | |
| Print shops, art rooms | | | | |
| Exercise rooms | | | | |
| Beauty salons | | | | |
| | | | | |
| | | | | |
| **Redecorating/Repair/Remodeling** | | | | |
| Emissions from new furnishings | | | | |
| Dust, fibers from demolition | | | | |
| Odors, volatile compounds | | | | |
| | | | | |
| | | | | |

## POLLUTANT AND SOURCE INVENTORY       Page _ of _

Building: _________________________________ File Number: _________

Completed by: _________________ Title: ______ Date Checked: _______

Using the list of potential source categories below, record any indications of contamination or suspected pollutants that may require further investigation or treatment. Sources of contamination may be constant or intermittent or may be linked to single, unrepeated events. For intermittent sources, indicate the time of peak activity or contaminant production, including correlations with weather (e.g., wind direction).

| Source Category | Checked | Needs Attention | Location | Comments |
|---|---|---|---|---|
|  |  |  |  |  |
|  |  |  |  |  |
|  |  |  |  |  |
|  |  |  |  |  |
|  |  |  |  |  |
|  |  |  |  |  |
|  |  |  |  |  |
|  |  |  |  |  |
|  |  |  |  |  |
|  |  |  |  |  |
|  |  |  |  |  |

## CHEMICAL INVENTORY

Building: _________________________________ File Number: _________

Address: _______________________________________________________

Completed by: _________________ Title: ______ Date Checked: _______

The inventory should include chemicals stored or used in the building for cleaning, maintenance, operations, and pest control. If there is an MSDS (Material Safety Data Sheet) for the chemical, put a check mark in the right-hand column. If not, ask the chemical supplier to provide the MSDS, if one is available.

| Date | Chemical/ Brand Name | Use | Storage Location(s) | MSDS on file? |
|---|---|---|---|---|
|  |  |  |  |  |
|  |  |  |  |  |
|  |  |  |  |  |
|  |  |  |  |  |
|  |  |  |  |  |
|  |  |  |  |  |
|  |  |  |  |  |
|  |  |  |  |  |
|  |  |  |  |  |
|  |  |  |  |  |
|  |  |  |  |  |
|  |  |  |  |  |

## HYPOTHESIS FORM                                    **Page 1 of 2**

Building: _________________________________ File Number: __________

Address: _________________________________________________________

Completed by: ____________________________________________________

**Complaint Area** (may be revised as the investigation progresses):

_________________________________________________________________

_________________________________________________________________

**Complaints** (e.g., summarize patterns of timing, location, number of people affected):

_________________________________________________________________

_________________________________________________________________

**HVAC**: Does the ventilation system appear to provide adequate outdoor air, efficiently distributed to meet occupant needs in the compliant area? If not, note potential problems.

_________________________________________________________________

_________________________________________________________________

Is there any apparent pattern connecting the location and timing of complaints with the HVAC system layout, condition or operating schedule?

_________________________________________________________________

**Pathways**: What pathways and driving forces connect the complaint area to locations of potential sources?

_________________________________________________________________

_________________________________________________________________

Are the flows opposite to those intended in the design? ____________

**Sources**: What potential sources have been identified in the complaint area or in locations associated with the complaint area (connected by pathways)?

_________________________________________________________________

_________________________________________________________________

Is the pattern of complaints consistent with any of these resources?____

## HYPOTHESIS FORM                                    **Page 2 of 2**

**Hypothesis**: Using the information gathered, determine the explanation for the problem.

______________________________________________________________

______________________________________________________________

______________________________________________________________

**Hypothesis testing**: How can this hypothesis be tested?

______________________________________________________________

______________________________________________________________

______________________________________________________________

If measurements have been taken, are the measurement results consistent with this hypothesis?

______________________________________________________________

______________________________________________________________

______________________________________________________________

**Results of Hypothesis Testing:**

______________________________________________________________

______________________________________________________________

______________________________________________________________

**Additional Information Needed:**

______________________________________________________________

______________________________________________________________

______________________________________________________________

**Source:**
United States Environmental Protection Agency
Office of Air and Radiation
Office if Atmospheric and Indoor Air Program
Indoor Air Division

U.S. Department of Health and Human Services
Public Health Service
Centers for Disease Control
National Institute for Occupational Safety and Health
Building Air Quality: *A Guide for Building Owners and Facility Managers.*
        1991.

# Appendix V

# *Indoor Air Quality (IAQ) Glossary of Terms*

Terms not defined herein should have their ordinary meaning within the context of their use. Ordinary meaning is as defined in "Webster's Ninth New Collegiate Dictionary."

**ACID AEROSOL:** Acidic liquid or solid particles that are small enough to become airborne. High concentrations of acid aerosols can be irritating to the lungs and have been associated with some respiratory disasters, such as asthma.

**ACTION LEVEL:** A term used to identify the level of indoor radon at which remedial action is recommended. (EPA's current action level is 4 pCi/L.)

**AHU:** See "Air Handling Unit."

**AIR CLEANING:** An IAQ control strategy to remove various airborne particulates and/or gases from the air. The three types of air cleaning most commonly used are particulate filtration, electrostatic precipitation, and gas sorption.

**AIR EXCHANGE RATE:** The rate at which outside air replaces indoor air in a space. Air exchange rate is expressed in one of two ways: the number of changes of outside air per unit of time air changes per hour (ACH); or the rate at which a volume of outside air enters per unit of time—cubic feet per minute (cfm).

**AIR HANDLING UNIT (AHU):** Includes a blower or fan, heating and/or cooling coils, and related equipment such as controls, condensate drain pans, and air filters. It does not include ductwork, registers or grilles, or boilers and chillers.

**AIR PASSAGES**: Openings through or within walls, through floors and ceilings, and around chimney flues and plumbing chases, that permit air to move out of the conditioned spaces of the building.

**ANIMAL DANDER**: Tiny scales of animal skin.

**ALLERGEN**: A substance capable of causing an allergic reaction because of an individual's sensitivity to that substance.

**ALLERGIC RHINITIS**: Inflammation of the mucous membranes in the nose that is caused by an allergic reaction.

**ANTIMICROBIAL**: Agents that kill microbial growth. See "disinfectant," "sanitizer," and "sterilizer."

**BIOLOGICAL CONTAMINANTS**: Agents derived from, or that are, living organisms (e.g., viruses, bacteria, fungi, and mammal and bird antigens) that can be inhaled and can cause many types of health effects including allergic reactions, respiratory disorders, hypersensitivity diseases, and infectious diseases. Also referred to as "micro-biologicals" or "microbials."

**BREATHING ZONE**: Area of a room in which occupants breathe as they stand, sit, or lie down.

**BUILDING ENVELOPE**: Elements of the building, including all external building materials, windows, and walls, that enclose the internal space.

**BUILDING-RELATED ILLNESS (BRI)**: Diagnosable illness whose symptoms can be identified and whose cause can be directly attributed to airborne building pollutants (e.g., Legionnaires disease, hypersensitivity pneumonitis). Also: A discrete, identifiable disease or illness than can be traced to a specific pollutant or source within a building. (Contrast with "Sick Building Syndrome").

**CEILING PLENUM**: Space below the flooring and above the suspended ceiling that accommodates the mechanical and electrical equipment and that is used as part of the air distribution system. The space is kept under negative pressure.

**CENTRAL AIR HANDLING UNIT (Central AHU)**: This is the

same as an Air Handling Unit, but serves more than one area.

**CFM**: Cubic feet per minute. The amount of air, in cubic feet, that flows through a given space in one minute. 1 CFM equals approximately 2 liters per second (l/s).

**CHEMICAL SENSITIZATION**: Evidence suggests that some people may develop health problems characterized by effects such as dizziness, eye and throat irritation, chest tightness, and nasal congestion that appear whenever they are exposed to certain chemicals. People may react to even trace amounts of chemicals to which they have become "sensitized."

**CO**: Carbon monoxide.

**$CO_2$**: Carbon dioxide.

**COMBINATION FOUNDATIONS**: Buildings constructed with more than one foundation type; e.g., basement/crawlspace or basement/slab-on-grade.

**COMMISSIONING**: Start-up of a building that includes testing and adjusting HVAC, electrical, plumbing, and other systems to assure proper functioning and adherence to design criteria. Commissioning also includes the instruction of building representatives in the use of the building systems.

**CONDITIONED AIR**: Air that has been heated, cooled, humidified, or dehumidified to maintain an interior space within the "comfort zone." (Sometimes referred to as "tempered" air.)

**CONSTANT AIR VOLUME SYSTEMS**: Air handling system that provides a constant air flow while varying the temperature to meet heating and cooling needs.

**DAMPERS**: Controls that vary airflow through an air outlet, inlet, or duct. A damper position may be immovable, manually adjustable or part of an automated control system.

**DIFFUSERS AND GRILLES**: Components of the ventilation system that distribute and return air to promote air circulation in the occupied space. As used here, supply air enters a space through a diffuser or vent and return air leaves a space through a grille.

**DISINFECTANTS**: One of the three groups of anti-microbials registered by EPA for public health uses.

**DRAIN TILE LOOP**: A continuous length of drain tile or perforated pipe extending around all or part of the internal or external perimeter of a basement or crawlspace footing.

**DRAIN TRAP**: A dip in the drain pipe of sinks, toilets, floor drains, etc., which is designed to stay filled with water, thereby preventing sewer gases from escaping into the room.

**ENVIRONMENTAL AGENTS**: Conditions other than indoor air contaminants that cause stress, comfort, and/or health problems (e.g., humidity extremes, drafts, lack of air circulation, noise, and over-crowding).

**ENVIRONMENTAL TOBACCO SMOKE (ETS)**: Mixture of smoke from the burning end of a cigarette, pipe, or cigar and smoke exhaled by the smoker (also secondhand smoke (SHS) or passive smoking).

**ERGONOMICS**: Applied science that investigates the impact of people's physical environment on their health and comfort (e.g., determining the proper chair height for computer operators).

**EXHAUST VENTILATION**: Mechanical removal of air from a portion of a building (e.g., piece of equipment, room, or general area).

**FLOW HOOD**: Device that easily measures airflow quantity, typically up to 2,500 cfm.

**FUNGI**: Any group of parasitic lower plants that lack chlorophyll, including molds and mildews.

**GAS SORPTION**: Devices used to reduce levels of airborne gaseous compounds by passing the air through materials that extract the gases. The performance of solid sorbents is dependent on the airflow rate, concentration of the pollutants, presence of other gases or vapors, and other factors.

**GOVERNMENTAL**: In the case of building codes, these are the State or local organizations/agencies responsible for building code enforcement.

**HEPA**: High efficiency particulate arrestance (filters).

**HUMIDIFIER FEVER**: A respiratory illness caused by exposure to toxins from microorganisms found in wet or moist areas in humidifiers and air conditioners. It is also called air conditioner or ventilation fever.

**HVAC**: Heating, ventilation, and air-conditioning system.

**HYPERSENSITIVITY DISEASES**: Diseases characterized by allergic responses to pollutants. The hypersensitivity diseases most clearly associated with indoor air quality are asthma, rhinitis, and hypersensitivity pneumonitis. Hypersensitivity pneumonitis is a rare but serious disease that involves progressive lung damage as long as there is exposure to the causative agent.

**HYPERSENSITIVITY PNEUMONITIS**: A group of respiratory diseases that cause inflammation of the lung (specifically granulomatous cells). Most forms of hypersensitivity pneumonitis are caused by the inhalation of organic dusts, including molds.

**IPM**: Integrated pest management.

**INDICATOR COMPOUNDS**: Chemical compounds, such as carbon dioxide, whose presence at certain concentrations may be used to estimate certain building conditions (e.g., airflow, presence of sources.).

**INDOOR AIR POLLUTANT**: Particles and dust, fibers, mists, bio-aerosols, and gases or vapors.

**MAKE-UP AIR**: See "Outdoor Air Supply."

**MAP OF RADON ZONES**: A U.S. EPA publication depicting areas of differing radon potential in both map form and in state specific booklets.

**MCS**: See "Multiple Chemical Sensitivity."

**MECHANICALLY VENTILATED CRAWLSPACE SYSTEM**: A system designed to increase ventilation within a crawlspace, achieve higher air pressure in the crawlspace relative to air pressure in the soil beneath the crawlspace, or achieve lower air pressure in the crawlspace relative to air pressure in the living spaces, by use of a fan.

**MICROBIOLOGICALS**: See "Biological Contaminants."

**MODEL BUILDING CODES**: The building codes published by the 4 Model Code Organizations and commonly adopted by state or other jurisdictions to control local construction activity.

**MODEL CODE ORGANIZATIONS**: Includes the following

agencies and the model building codes they promulgate:

- Building Officials and Code Administrators International, Inc. (BOCA National Building Code/1993 and BOCA National Mechanical Code/1993);
- International Conference of Building Officials (Uniform Building Code/1991 and Uniform Mechanical Code/1991);
- Southern Building Code Congress, International, Inc. (Standard Building Code/1991 and Standard Mechanical Code/1991);
- Council of American Building Officials (CABO One- and Two-Family Dwelling Code/1992 and CABO Model Energy Code/1993).

**MULTIPLE CHEMICAL SENSITIVITY (MCS)**: A condition in which a person reports sensitivity or intolerance (as distinct from "allergic") to a number of chemicals and other irritants at very low concentrations. There are different views among medical professionals about the existence, causes, diagnosis, and treatment of this condition.

**NEGATIVE PRESSURE**: Condition that exists when less air is supplied to a space than is exhausted from the space, so the air pressure within that space is less than that in surrounding areas. Under this condition, if an opening exists, air will flow from surrounding areas into the negatively pressurized space.

**ORGANIC COMPOUNDS**: Chemicals that contain carbon. Volatile organic compounds vaporize at room temperature and pressure. They are found in many indoor sources, including many common household products and building materials.

**OUTDOOR AIR SUPPLY**: Air brought into a building from the outdoors (often through the ventilation system) that has not been previously circulated through the system. It is also known as "Make-Up Air."

**PELs**: Permissible Exposure Limits (standards set by the Occupational, Safety and Health Administration—OSHA).

**PICOCURIE (pCi)**: A unit for measuring radioactivity often expressed as picocuries per liter (pCi/L) of air.

**PLENUM**: Air compartment connected to a duct or ducts.

**PM**: Preventive Maintenance.

**POLLUTANT PATHWAYS**: Avenues for distribution of pollutants in a building. HVAC systems are the primary pathways in most buildings; however all building components interact to affect how air movement distributes pollutants.

**POSITIVE PRESSURE**: Condition that exists when more air is supplied to a space than is exhausted, so the air pressure within that space is greater than that in surrounding areas. Under this condition, if an opening exists, air will flow from the positively pressurized space into surrounding areas.

**PPM**: Parts per million.

**PRESSED WOOD PRODUCTS**: A group of materials used in building and furniture construction that are made from wood veneers, particles, or fibers bonded together with an adhesive under heat and pressure.

**PRESSURE, STATIC**: In flowing air, the total pressure minus velocity pressure. It is the portion of the pressure that pushes equally in all directions.

**PRESSURE, TOTAL**: In flowing air, the sum of the static pressure and the velocity pressure.

**PRESSURE, VELOCITY**: In flowing air, the pressure due to the velocity and density of the air.

**PREVENTIVE MAINTENANCE**: Regular and systematic inspection, cleaning, and replacement of worn parts, materials, and systems. Preventive maintenance helps to prevent parts, material, and systems failure by ensuring that parts, materials and systems are in good working order.

**PSYCHOGENIC ILLNESS**: This syndrome has been defined as a group of symptoms that develop in an individual (or a group of individuals in the same indoor environment) who are under some type of physical or emotional stress. This does not mean that individuals have a psychiatric disorder or that they are imagining symptoms.

**PSYCHOSOCIAL FACTORS**: Psychological, organizational, and personal stressors that could produce symptoms similar to those caused by poor indoor air quality.

**RADIANT HEAT TRANSFER**: Radiant heat transfer occurs when there is a large difference between the temperatures of two surfaces that are exposed to each other, but are not touching.

**RADON (Rn) AND RADON DECAY PRODUCTS**: Radon is a radioactive gas formed in the decay of uranium. The radon decay products (also called radon daughters or progeny) can be breathed into the lung where they continue to release radiation as they further decay.

**RE-ENTRAINMENT**: Situation that occurs when the air being exhausted from a building is immediately brought back into the system through the air intake and other openings in the building envelope.

**RE-ENTRY**: Situation that occurs when the air being exhausted from a building is immediately brought back into the system through the air intake and other openings in the building envelope.

**RELs**: Recommended Exposure Limits (recommendations made by the National Institute for Occupational Safety and Health (NIOSH).

**SANITIZER**: One of three groups of anti-microbials registered by EPA for public health uses. EPA considers an anti-microbial to be a sanitizer when it reduces but does not necessarily eliminate all the microorganisms on a treated surface. To be a registered sanitizer, the test results for a product must show a reduction of at least 99.9% in the number of each test microorganism over the parallel control.

**SHORT-CIRCUITING**: A situation that occurs when the supply air flows to return or exhaust grilles before entering the breathing zone (area of a room where people are). To avoid short-circuiting, the supply air must be delivered at a temperature and velocity that results in mixing throughout the space.

**SICK BUILDING SYNDROME (SBS)**: Term that refers to a set of symptoms that affect some number of building occupants during the time they spend in the building and diminish or go away during periods when they leave the building. Cannot be traced to specific pollutants or sources within the building. (Contrast with "Building related illness").

**SOIL GAS**: The gas present in soil which may contain radon.

**SOIL-GAS-RETARDER**: A continuous membrane or other comparable material used to retard the flow of soil gases into a building.

**SOURCES**: Sources of indoor air pollutants. Indoor air pollutants can originate within the building or be drawn in from outdoors. Common sources include people, room furnishings such as carpeting, photocopiers, art supplies, etc.

**STACK EFFECT**: The overall upward movement of air inside a building that results from heated air rising and escaping through openings in the building super structure, thus causing an indoor pressure level lower than that in the soil gas beneath or surrounding the building foundation.

**STATIC PRESSURE**: A condition that exists when an equal amount of air is supplied to and exhausted from a space. At static pressure, equilibrium has been reached.

**STERILIZER**: One of three groups of anti-microbials registered by EPA for public health uses. EPA considers an anti-microbial to be a sterilizer when it destroys or eliminates all forms of bacteria, fungi, viruses, and their spores. Because spores are considered the most difficult form of a microorganism to destroy, EPA considers the term sporicide to be synonymous with "sterilizer."

**SUB-SLAB DEPRESSURIZATION SYSTEM (ACTIVE)**: A system designed to achieve lower sub-slab air pressure relative to indoor air pressure by use of a fan-powered vent drawing air from beneath the slab.

**SUB-SLAB DEPRESSURIZATION SYSTEM (PASSIVE)**: A system designed to achieve lower sub-slab air pressure relative to indoor air pressure by use of a vent pipe routed through the conditioned space of a building and connecting the sub-slab areas with outdoor air, thereby relying solely on the convective flow of air upward in the vent to draw air from beneath the slab.

**SUB-MEMBRANE DEPRESSURIZATION SYSTEM**: A system designed to achieve lower sub-membrane air pressure relative to crawlspace air pressure by use of a fan-powered vent drawing air from under the soil-gas-retarder membrane.

**TRACER GASES**: Compounds, such as sulfur hexaflouride, which are used to identify suspected pollutant pathways and to quantify ventilation rates. Trace gases may be detected qualitatively by their odor or quantitatively by air monitoring equipment.

**TLVs**: Threshold Limit Values (guidelines recommended by the American Conference of Governmental Industrial Hygienists).

**TVOCs**: Total volatile organic compounds. See "Volatile Organic Compounds (VOCs)."

**UNIT VENTIALATOR**: A fan-coil unit package device for applications in which the use of outdoor- and return-air mixing is intended to satisfy tempering requirements and ventilation needs.

**VARIABLE AIR VOLUME SYSTEM (VAV)**: Air handling system that conditions the air to constant temperature and varies the outside airflow to ensure thermal comfort.

**VENTILATION AIR**: Defined as the total air, which is a combination of the air brought inside from outdoors and the air that is being re-circulated within the building. Sometimes, however, used in reference only to the air brought into the system from the outdoors; the EPA defines this air as "outdoor air ventilation."

**VENTILATION RATE**: The rate at which indoor air enters and leaves a building. This rate is expressed in one of two ways: the number of changes of outdoor air per unit of time (air changes per hour, or "ach") or the rate at which a volume of outdoor air enters per unit of time (cubic feet per minute, or "cfm").

**VOLATILE ORGANIC COMPOUNDS (VOCs)**: Compounds that vaporize (become a gas) at room temperature. Common sources which may emit VOCs into indoor air include housekeeping and maintenance products, and building and furnishing materials. In sufficient quantities, VOCs can cause eye, nose, and throat irritations, headaches, dizziness, visual disorders, memory impairment; some are known to cause cancer in animals; some are suspected of causing or are known to cause, cancer in humans. At present, not much is known about what health effects occur at the levels of VOCs typically found in public and commercial buildings.

**ZONE**: The occupied space or group of spaces within a building which has its heating or cooling controlled by a single thermostat.

**Source**: United States Environmental Protection Agency

# *Appendix VI*
# *Accessible Furniture*

Workplace accessibility furnishings. (Courtesy of KI, 1330 Bellevue St., Green Bay, WI 54302. Ph. 800-424-2432, or 920-468-8100; fax 920-468-0280.)

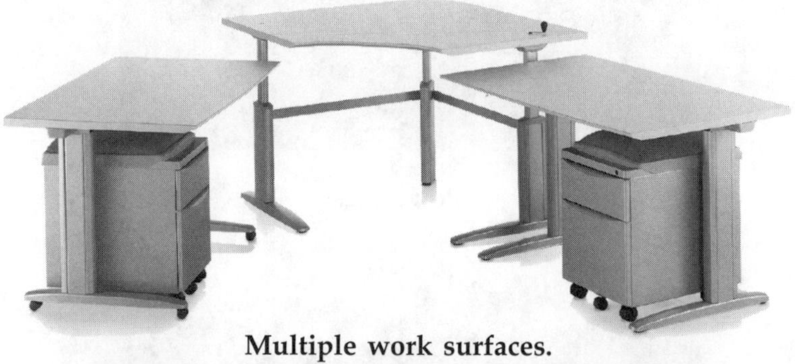

Multiple work surfaces.

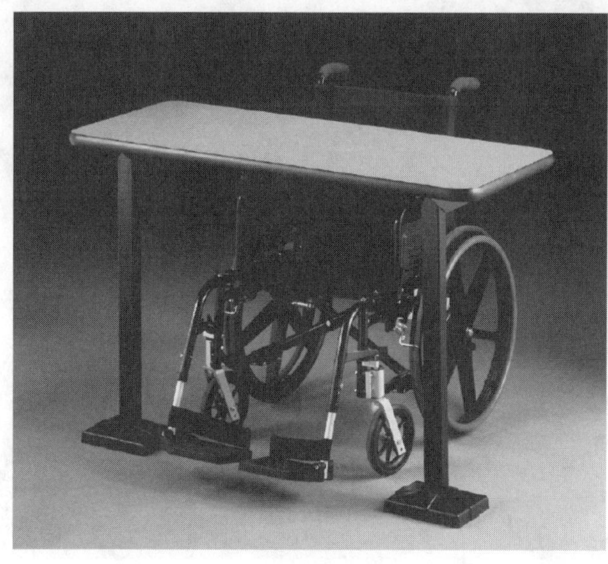

Fixed seating.

Powercom.

Sitting flat.

Adjustable table.

Drawer pull.

Dual height adjustable work surface.

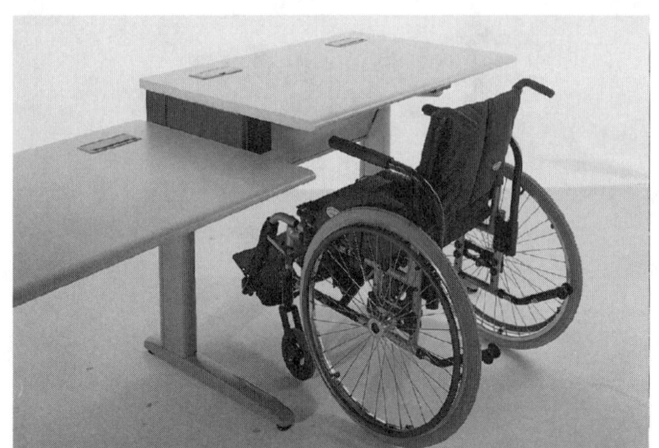

Tandem desktop
wheelchair.

Wheelchair table.

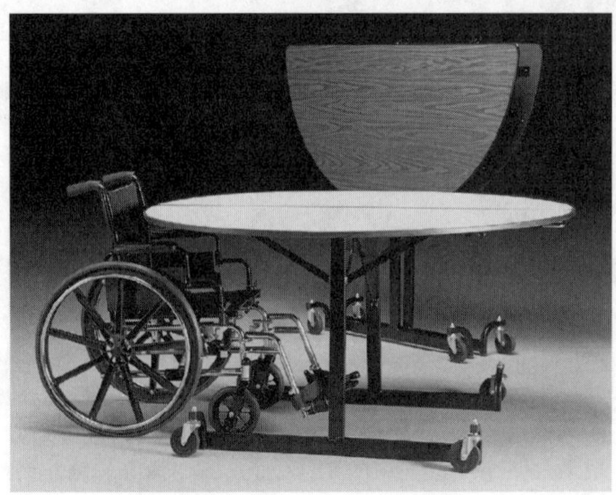

Wheelchair tables.

# Index